Naturally Occurring Benzodiazepines, Endozepines, and their Receptors

Frontiers in Neurotherapeutics Series

Series Editor:
Diana Amantea
Laura Berliocchi
Rossella Russo

ABOUT THE SERIES

New contexts are needed for the discussion and integration of basic and clinical data in order to translate the dynamic progress in the field of neuroscience into more effective therapies and patients' health improvements.

The Frontiers in Neurotherapeutics Series provides a multidisciplinary platform to discuss innovative, methodological, and pharmacological approaches aimed at the identification and validation of novel neurotherapeutics. It features reviews and original articles from basic and translational research on cutting-edge topics in diverse and integrated areas of neuroscience, including neuro-immunology and neuro-oncology.

Recent titles in this series:

Naturally Occurring Benzodiazepines, Endozepines, and their Receptors
Implications for Benzodiazepine Therapy and Withdrawal
Robert B. Raffa and Diana Amantea

Rational Basis for Clinical Translation in Stroke Therapy
Giuseppe Micieli and Diana Amantea

Drug Repositioning
Approaches and Applications for Neurotherapeutics
Joel Dudley and Laura Berliocchi

Mapping of Nervous System Diseases via MicroRNAs
Christian Barbato and Francesca Ruberti

Naturally Occurring Benzodiazepines, Endozepines, and their Receptors

Implications for Benzodiazepine Therapy and Withdrawal

Edited by

Robert B. Raffa and Diana Amantea

CRC Press
Taylor & Francis Group
Boca Raton London New York

CRC Press is an imprint of the
Taylor & Francis Group, an informa business

First edition published 2022
by CRC Press
6000 Broken Sound Parkway NW, Suite 300, Boca Raton, FL 33487-2742

and by CRC Press
2 Park Square, Milton Park, Abingdon, Oxon, OX14 4RN

CRC Press is an imprint of Taylor & Francis Group, LLC

Library of Congress Cataloging-in-Publication Data
Names: Raffa, Robert B., editor. | Amantea, Diana, editor.
Title: Naturally occurring benzodiazepines, endozepines, and their
receptors : implications for benzodiazepine therapy and withdrawal /
edited by Robert B. Raffa, Diana Amantea.
Description: First edition. | Boca Raton, FL : CRC Press, 2021. | Series:
Frontiers in neurotherapeutics | Includes
bibliographical references and index.
Identifiers: LCCN 2021021778 (print) | LCCN 2021021779 (ebook) |
ISBN 9780367409067 (paperback) | ISBN 9781032043104 (hardback) |
ISBN 9780367814373 (ebook)
Subjects: LCSH: Benzodiazepines—Physiological effect. |
Benzodiazepines—Receptors.
Classification: LCC QP801.B53 N38 2021 (print) | LCC QP801.B53 (ebook) |
DDC 615.7/8—dc23 LC record available at
https://lccn.loc.gov/2021021778
LC ebook record available at
https://lccn.loc.gov/2021021779

ISBN: 978-0-367-40906-7 (pbk)
ISBN: 978-1-032-04310-4 (hbk)
ISBN: 978-0-367-81437-3 (ebk)

Typeset in Times
by codeMantra

Contents

Preface

The first benzodiazepine to enter medical practice, chlordiazepoxide, was discovered in 1957 and was first marketed in 1960 as Librium®. Many other benzodiazepines followed; there are now more than a dozen on the market. Benzodiazepines presented distinct pharmacologic and clinical advantages over extant therapy such as barbiturates, which were less specific and were fraught with a plethora of adverse effects and other problems. The better specificity and distinct adverse-effect advantages of the benzodiazepines over the barbiturates rightfully catapulted the benzodiazepines to widespread use and commercial success. Unfortunately, a consequence of a familiarity and success with appropriate and circumspect prescribing of benzodiazepines was a transition into good-intentioned, but ever-expanding (over)use for unproven applications, and for durations that greatly exceeded clinical trial experience, product labeling, and proof of efficacy and/or safety. Particularly of concern, the long-term use was naïvely assumed to have the same safety as (sub)acute use, in the absence of supporting data. The trend was encouraged, if not overtly led, by aggressive and questionable advertising. In retrospect, these were highly unfortunate assumptions and marketing practices. The serious problems of improper benzodiazepine prescribing, and even the withdrawal symptoms, were described only one year after marketing. Now other problems are coming to the fore, long known to patients but under-recognized, overlooked, or ignored by healthcare providers, as highlighted in the recent special communication by the United States FDA. Among the problems is the existence of mechanistically perplexing protracted withdrawal symptoms experienced by some patients after discontinuation of therapy. It is clear that when they are properly prescribed and used, positive allosteric modulators of GABA-A receptors such as the benzodiazepines and Z-drugs are usually safe and effective therapy for most patients when used for a short period of time, *e.g.*, 2–4 weeks. However, a subset of patients experience a lack of efficacy, severe adverse effects, and/or protracted withdrawal symptoms, which can be independent of dose and of the duration of use. Healthcare providers should become aware of and acknowledge these shortcomings, and take them seriously. Unfortunately, there is currently no good way to understand some of the phenomena, or to predict outcome prior to treatment. It is hoped that this review of endogenous benzodiazepines sheds some light on the topic.

Diana Amantea, PhD
Rende (Cosenza), Italy

Robert B. Raffa, PhD
Tucson, AZ, USA

January 2021

About the Editors

Robert B. Raffa is Adjunct Professor at the University of Arizona College of Pharmacy and Professor Emeritus at Temple University School of Pharmacy. He earned bachelor's degrees in Chemical Engineering and in Physiological Psychology (both from the University of Delaware), master's degrees in Biomedical Engineering (from Drexel University) and Toxicology (from Thomas Jefferson University), and a doctorate in Pharmacology (from Temple University School of Medicine). He was a Research Fellow and a Team Co-Leader for drug discovery at Johnson & Johnson. He is a cofounder of CaRafe Drug Innovation and Enalare Therapeutics, and is CSO of Neumentum, Inc. He is co-holder of several patents and has published more than 350 papers. He has co-authored or edited several books on pharmacology and thermodynamics, is a Co-Editor of two journals, is a past president of the Mid-Atlantic Pharmacology Society, and is the recipient of research and teaching awards.

Diana Amantea is Associate Professor of Pharmacology at the Department of Pharmacy, Health and Nutritional Sciences of the University of Calabria (Italy), where she is the leader of the Stroke Research Unit at the Section of Preclinical and Translational Pharmacology operating in the frame of the Italian Stroke Organization (ISO) Basic Science. She earned a PharmD cum laude from the University of Calabria in 1998 and a PhD in Pharmacology from the University of Birmingham (UK) in 2003. She continued her postdoctoral activity at the Department of Pharmacology of the Medical School of Birmingham University where, under the guidance of Prof. Norman G. Bowery, she worked on the involvement of GABAB receptors in neurological disorders. During her formative years, she was visiting researcher at respected research laboratories of the Universities of Madrid, Lausanne, Florence, and Rome. She has published more than 60 papers in peer-reviewed journals. She is a member of the Editorial Board and Guest Editor of the 2016 Neuroscience section of *Current Opinion in Pharmacology* (Elsevier), and the founder and editor of *Frontiers in Neurotherapeutics* series (CRC Press).

Contributors

Diana Amantea
Department of Pharmacy, Health and Nutritional Sciences
University of Calabria
Rende (CS), Italy

Rafael Coveñas
Instituto de Neurosciencias de Castilla y León (INCYL)
Laboratorio de Neuroanatomía de los Sistemas Peptidérgicos
Salamanca, Spain

University of Salamanca
Salamanca, Spain

Monica Gallo
Department of Molecular Medicine and Medical Biotechnology
University of Naples Federico II
Naples, Italy

Antonio Garofalo
Department of Pharmacy, Health and Nutritional Sciences
University of Calabria
Rende (CS), Italy

Fedora Grande
Department of Pharmacy, Health and Nutritional Sciences
University of Calabria
Rende (CS), Italy

Tina Hinton
Pharmacology, Sydney Pharmacy School, Faculty of Medicine and Health Sciences
The University of Sydney
Sydney, Australia

Graham A. R. Johnston
Pharmacology, Sydney Pharmacy School, Faculty of Medicine and Health Sciences
The University of Sydney
Sydney, Australia

Jan M. Kitzen
Kitzen Pharmaceutical Consulting
Collegeville, PA

Jo Ann LeQuang
NEMA Research, Inc.
Naples, FL

Domenico Montesano
Department of Pharmaceutical Sciences, Section of Food Sciences and Nutrition
University of Perugia
Perugia, Italy

Department of Pharmacy
University of Naples Federico II
Naples, Italy

Maria Antonietta Occhiuzzi
Department of Pharmacy, Health and Nutritional Sciences
University of Calabria
Rende (CS), Italy

Michael H. Ossipov
Department of Pharmacology (Emeritus), and Comprehensive Pain and Addiction Center
University of Arizona College of Medicine
Tucson, AZ

Oné R. Pagán
Department of Biology
West Chester University
West Chester, PA

John F. Peppin
Marian University College of Osteopathic Medicine (Adjunct)
Indianapolis, IN

Pikeville University College of Osteopathic Medicine
Pikeville, KY

Robert B. Raffa
Temple University School of Pharmacy (Emeritus)
Philadelphia, PA

University of Arizona College of Pharmacy (Adjunct)
Tucson, AZ

Neumentum, Inc.
Morristown, NJ

Enalare Therapeutics
Princeton, NJ, USA

Felix-Martin Werner
Höhere Berufsfachschule für Altenpflege und Ergotherapie der Euro Akademie Pößneck
Pößneck, Germany

Instituto de Neurosciencias de Castilla y León (INCYL), Laboratorio de Neuroanatomía de los Sistemas Peptidérgicos
Salamanca, Spain

University of Salamanca
Salamanca, Spain

Section I

Introduction and Basic Principles

1 Introduction

Robert B. Raffa
Temple University School of Pharmacy (Emeritus)
University of Arizona College of Pharmacy (Adjunct)
Neumentum, Inc.
Enalare Therapeutics

CONTENTS

ENDOGENOUS OPIOIDS LEAD THE WAY

The catalyst for the train-of-thought that led to endogenous benzodiazepines was the discovery of endogenous opioids. In the beginning, there was only the serendipitous recognition of the therapeutic (and other) effects produced by plant substances found in the liquid that exudes from the notched unripe capsule of the opium poppy (*opos* from the Greek word meaning juice). This level of knowledge sufficed for thousands of years. Then, a large advance in understanding occurred with the now famous publication by Sertürner in 1806[1] of the isolation of a major active ingredient of the opium poppy—morphine—and identification of its chemical structure (codeine was isolated a few years later). This put the drug-effect sequence on a firm molecular chemical basis. And it also raised a new question: how could a small molecule be responsible for a whole-body effect? And another question gradually arose: why do some chemical structures mimic morphine and codeine, but others do not? Specifically, why was there a strict relationship between chemical structure and function (SAR)? And more specifically, why was there a strict relationship between three-dimensional chemical structure and function? Why, for example, is one enantiomer of a racemic mixture often manyfold more potent in producing an effect than is the corresponding other enantiomer?

Then came a real surprise. Weijlard and Erikson in 1942[2] reported that a compound that had a structure somewhat similar to morphine and codeine ("opiates") did not produce similar effects, but could *reverse* them. That is, it acted like an antagonist would at a receptor that mediated the effects of an agonist. So by extension, morphine-induced effects seemed to be mediated by a receptor—and morphine acted by binding to (have affinity for), and activating (intrinsic activity or efficacy), that receptor. Years of accumulated data, development of more advanced analytical techniques, and increasing refinement of in vivo studies pointed to the existence of not only one opioid receptor, but several (i.e., the existence of subtypes of opioid receptor). Proof came with the suggestion to use radiolabeled compounds in what is now known as "radioligand binding studies,"[3] and the successful use of this technique was employed by three groups and reported almost simultaneously in 1973.[4–6] There were now opioid receptors. The next obvious question was, "Why are they there?" Could there be endogenous opiate-like substances that interacted with these receptors?

Then, Akil and colleagues[7] reported that foot-shock stress to animals produced analgesia. How was this possible? The fact that an opioid receptor antagonist (naloxone) reversed the effect of foot-shock stress suggested the answer—the stress must have elicited the release of some opiate-like endogenous substance(s). This would explain the reversal by an opiate receptor antagonist. It was not long before endogenous opiate-like peptides (termed "opioids") were identified, such as the endorphins, enkephalins, deltorphins, dynorphin, and more.[8–12] The demonstration that the inhibition of metabolic degradation of these endogenous substances produced opiate-like effects such as analgesia completed the circle by establishing a physiological role for the endogenous opioids.[13]

A similar story evolved for the benzodiazepines, but it followed by about 20 years later.[14]

ENDOGENOUS BENZODIAZEPINES

In a like manner to the opioids, benzodiazepines were used clinically before their site of action was known. The first chemical synthesis of benzodiazepines was accomplished in the mid-1950s, and the first one to market, chlordiazepoxide (LIBRIUM), appeared in 1960.[15] It was subsequently determined that the effects of benzodiazepines (and the Z-drugs) are mediated by agonist action at specific receptors (named after the extant chemical class, the benzodiazepine receptor) located on neurons.[16,17] But unlike the arrangement of the opioid receptor, the benzodiazepine receptor is part of another receptor, the $GABA_A$ receptor complex.[18–20] The binding of benzodiazepine agonists allosterically modulates (enhances) the action of GABA (γ-aminobutyric acid)-induced transmembrane neuronal Cl^- influx, thereby increasing the transmembrane potential difference, rendering the neuron less likely to respond with an action potential in response to a stimulus. This results in an anxiolytic effect (substantiated by the opposite effect—seizure—produced by *inverse* benzodiazepine receptor agonists). But it was only in 1986 that benzodiazepines were isolated from the mammalian brain.[21] The intriguing rest of the story is told in this volume and in recent comprehensive reviews.[14,22,23]

IMPLICATIONS

GABA is the major inhibitory amino acid neurotransmitter in the brain. Too little or too great an influence by GABA disturbs the critical balance between GABA and the counterbalancing amino acid excitatory neurotransmitters (primarily glutamate and aspartate) (Figure 1.1).[24] Disturbance in GABA levels, or functioning, has been associated with anxiety.[25–28] Benzodiazepine therapy is predicated on the ability of benzodiazepine receptor agonists to modulate GABA action at the $GABA_A$ receptor complex (Figure 1.2).

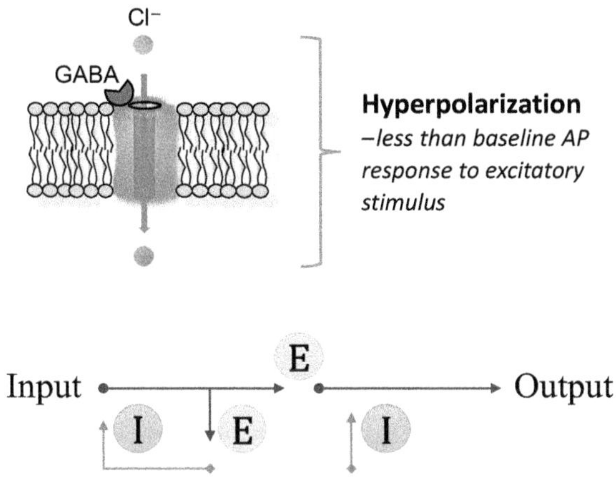

FIGURE 1.1 The binding of the inhibitory (I) neurotransmitter GABA to its site on the $GABA_A$ receptor complex "opens" the chloride ion channel, promoting a greater influx of Cl^- ions. The greater influx of Cl^- ions increases the neuron's transmembrane potential difference (hyperpolarization), which renders it less likely to reach threshold and "fire" (an action potential) in response to an excitatory (E) stimulus by a neurotransmitter such as aspartate.[24]

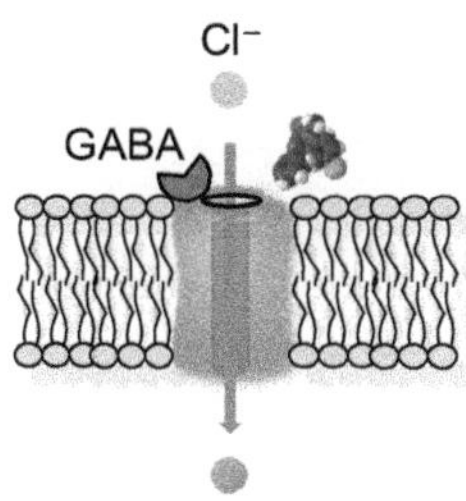

FIGURE 1.2 The benzodiazepine receptor binding site on the large $GABA_A$ receptor complex.

The magnitude of the clinical effect (e.g., anxiolytic) that is produced by the administration of a benzodiazepine is dependent on the dose administered. At the molecular level, this translates to the concentration (number) of benzodiazepine molecules at the receptor site, where molecules of benzodiazepine (B_z) drug reversibly combine with molecules of benzodiazepine receptor (R_B) to form a drug–receptor complex (B_zR_B), according to the law of mass action[29]:

$$B_Z + R_B \Leftrightarrow B_Z R_B \rightarrow\rightarrow\rightarrow \text{drug effect.}$$

This equation is a hallmark of pharmacodynamics, and is extremely helpful in understanding drug action and quantifying the relationship between the magnitude of drug dose and the magnitude of drug-induced effect. Indeed, this construct has led to many practical applications in basic and clinical pharmacology.[30,31]

However, it needs to be used in context. Drug (e.g., benzodiazepine) molecules are distributed to where their receptors are located (Figure 1.3). But if there is also

$$B_Z + R_B \leftrightarrow B_ZR_B$$

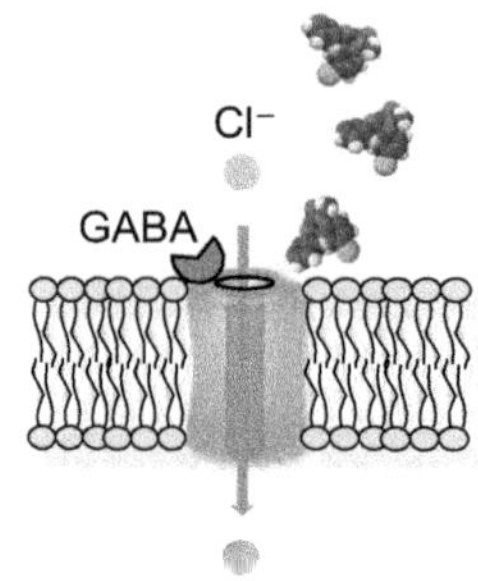

FIGURE 1.3 The binding of a ligand to its receptor follows the law of mass action. According to receptor occupancy theory, a greater concentration of agonist produces a greater effect.

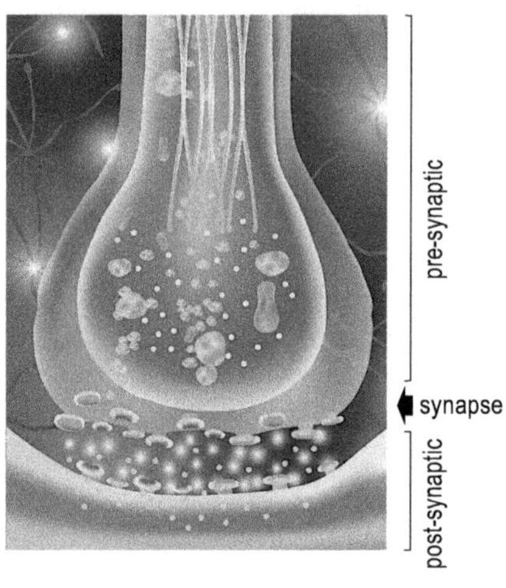

FIGURE 1.4 Synapse showing the presence of endogenous neurotransmitter substance. There is always some basal release of neurotransmitter from the presynaptic neuron into the synaptic cleft; therefore, exogenous drug molecules are always added to an existing amount. Modified from illustration purchased from iStock.

endogenous receptor substance there, then the amount of receptor occupancy is more accurately represented by the total amount of administered drug *plus endogenous substance* (Figure 1.4):

$$\left(B_{Z\text{-administerd}} + R_{Z\text{-endogenous}}\right) + R_B \Leftrightarrow B_ZR_B \rightarrow\rightarrow\rightarrow \text{drug effect.}$$

This changes everything, since:

- There is always a basal level of endogenous ligand ($B_{Z\text{-endogenous}}$) that is present
- The level of endogenous ligand ($B_{Z\text{-endogenous}}$) can differ among patients (polymorphisms in synthesis or degradation)

- The amount of endogenous ligand ($B_{Z\text{-endogenous}}$) present can change with age and aging
- The amount of endogenous ligand ($B_{Z\text{-endogenous}}$) present can change with diet and other drugs (OTC or prescription)
- The amount of endogenous ligand ($B_{Z\text{-endogenous}}$) present can change with disease
- The amount of endogenous ligand ($B_{Z\text{-endogenous}}$) present can change in response to pain conditions
- Mental status (e.g., anxiety, depression) can greatly influence the amount of endogenous ligand ($B_{Z\text{-endogenous}}$) present
- During therapy, compensatory (usually opposite) physiological responses occur, which become an important factor during cessation of therapy
- The amount of endogenous ligand ($B_{Z\text{-endogenous}}$) undeniably changes with the administration of an exogenous benzodiazepine (or Z-drug)

This means:

- A patient who is administered a benzodiazepine (or Z-drug) will respond to that drug in a unique manner, depending on the above conditions
- Their response will change with time (whether observed externally or not)
- The above conditions will uniquely affect the extent and magnitude of the adverse effects they experience
- The above conditions will affect the degree and duration of the symptoms experienced during cessation of therapy (withdrawal)

That is why the existence of endogenous benzodiazepines is so important. That is why it is not important whether the endogenous benzodiazepines arose by synthetic processes within the brain or periphery, or they are of exogenous origin. The impact on the clinical use of benzodiazepines is the same. And it is largely not appreciated.

REFERENCES

1. Krishnamurti C, Rao SC. The isolation of morphine by Serturner. *Indian J Anaesth* 2016;60:861–2.
2. Weijlard J, Erikson AE. N-allylnormorphine. *J Am Chem Soc* 1942;64: 869–70.
3. Goldstein A, Lowney LI, Pal BK. Stereospecific and nonspecific interactions of the morphine congener levorphanol in subcellular fractions of mouse brain. *Proc Natl Acad Sci U S A* 1971;68:1742–7.
4. Pert CB, Snyder SH. Opiate receptor: demonstration in nervous tissue. *Science* 1973;179:1011–4.
5. Simon EJ, Hiller JM, Edelman I. Stereospecific binding of the potent narcotic analgesic (3H) Etorphine to rat-brain homogenate. *Proc Natl Acad Sci U S A* 1973;70:1947–9.
6. Terenius L. Characteristics of the "receptor" for narcotic analgesics in synaptic plasma membrane fraction from rat brain. *Acta Pharmacol Toxicol (Copenh)* 1973;33:377–84.
7. Madden J, Akil H, Patrick RL, Barchas JD. Stress-induced parallel changes in central opioid levels and pain responsiveness in the rat. *Nature* 1977;265:358–60.
8. Kosterlitz HW, Waterfield AA. In vitro models in the study of structure-activity relationships of narcotic analgesics. *Annu Rev Pharmacol* 1975;15:29–47.

9. Hughes J, Smith TW, Kosterlitz HW, Fothergill LA, Morgan BA, Morris HR. Identification of two related pentapeptides from the brain with potent opiate agonist activity. *Nature* 1975;258:577–80.
10. Birdsall NJ, Bradbury AF, Burgen AS, Hulme EC, Smyth DG, Snell CR. Interactions of peptides derived from the C-fragment of beta-lipotropin with brain opiate receptors [proceedings]. *Br J Pharmacol* 1976;58:460P–1P.
11. Goldstein A, Fischli W, Lowney LI, Hunkapiller M, Hood L. Porcine pituitary dynorphin: complete amino acid sequence of the biologically active heptadecapeptide. *Proc Natl Acad Sci U S A* 1981;78:7219–23.
12. Erspamer V, Melchiorri P, Falconieri-Erspamer G, et al. Deltorphins: a family of naturally occurring peptides with high affinity and selectivity for delta opioid binding sites. *Proc Natl Acad Sci U S A* 1989;86:5188–92.
13. Poras H, Bonnard E, Dange E, Fournie-Zaluski MC, Roques BP. New orally active dual enkephalinase inhibitors (DENKIs) for central and peripheral pain treatment. *J Med Chem* 2014;57:5748–63.
14. Izquierdo I, Medina J. *Naturally Occurring Benzodiazepines*. New York: Ellis Horwood; 1993.
15. Lopez-Munoz F, Alamo C, Garcia-Garcia P. The discovery of chlordiazepoxide and the clinical introduction of benzodiazepines: half a century of anxiolytic drugs. *J Anxiety Disord* 2011;25:554–62.
16. Braestrup C, Nielsen M, Olsen CE. Urinary and brain beta-carboline-3-carboxylates as potent inhibitors of brain benzodiazepine receptors. *Proc Natl Acad Sci U S A* 1980;77:2288–92.
17. Mohler H, Okada T. Benzodiazepine receptor: demonstration in the central nervous system. *Science* 1977;198:849–51.
18. Olsen RW, Tobin AJ. Molecular biology of GABAA receptors. *FASEB J* 1990;4:1469–80.
19. Pritchett DB, Sontheimer H, Shivers BD, et al. Importance of a novel GABAA receptor subunit for benzodiazepine pharmacology. *Nature* 1989;338:582–5.
20. Schofield PR, Darlison MG, Fujita N, et al. Sequence and functional expression of the GABA A receptor shows a ligand-gated receptor super-family. *Nature* 1987;328:221–7.
21. De Blas AL, Sangameswaran L. Demonstration and purification of an endogenous benzodiazepine from the mammalian brain with a monoclonal antibody to benzodiazepines. *Life Sci* 1986;39:1927–36.
22. Farzampour Z, Reimer RJ, Huguenard J. Endozepines. *Adv Pharmacol* 2015;72:147–64.
23. Tonon MC, Vaudry H, Chuquet J, et al. Endozepines and their receptors: structure, functions and pathophysiological significance. *Pharmacol Ther* 2020;208:107386.
24. Ben-Ari Y. The GABA excitatory/inhibitory developmental sequence: a personal journey. *Neuroscience* 2014;279:187–219.
25. Nuss P. Anxiety disorders and GABA neurotransmission: a disturbance of modulation. *Neuropsychiatr Dis Treat* 2015;11:165–75.
26. Lydiard RB. The role of GABA in anxiety disorders. *J Clin Psychiatry* 2003;64 Suppl 3:21–7.
27. Mohler H. The GABA system in anxiety and depression and its therapeutic potential. *Neuropharmacology* 2012;62:42–53.
28. Babaev O, Piletti Chatain C, Krueger-Burg D. Inhibition in the amygdala anxiety circuitry. *Exp Mol Med* 2018;50:18.
29. Ferner RE, Aronson JK. Cato Guldberg and Peter Waage, the history of the law of mass action, and its relevance to clinical pharmacology. *Br J Clin Pharmacol* 2016;81:52–5.
30. Tallarida RJ. Interactions between drugs and occupied receptors. *Pharmacol Ther* 2007;113:197–209.
31. Raffa RB, Tallarida RJ. 'Affinity': historical development in chemistry and pharmacology. *Bull His Chem* 2010;35:7–16.

2 Benzodiazepines and Related Substances: Chemistry

Fedora Grande, Maria Antonietta Occhiuzzi and Antonio Garofalo
Health and Nutritional Sciences, University of Calabria

CONTENTS

INTRODUCTION

Benzodiazepines (BDZs) are bicyclic heterocyclic compounds structurally characterized by the fusion of a benzene with a diazepine, a seven-membered ring containing two nitrogen atoms. The identification of compounds belonging to this class of molecules, endowed with psychotropic activity, represents a peculiar example of successful drug discovery in modern medicinal chemistry (Kyburz, 1989). Most of these compounds find wide application in the management of several central nervous system (CNS) disorders. The clinical efficacy of most of the 1,4-BDZs derives from their ability to potentiate the effects of γ-aminobutyric acid (GABA), the major inhibitory neurotransmitter of the CNS, acting as positive allosteric modulators of $GABA_A$ (GABA subtype A receptor) (Maramai et al., 2020; McKernan et al., 2000). In fact, after binding to a specific modulatory/allosteric site of this receptor, BDZs enhance GABA-mediated neuron hyperpolarization by increasing the opening frequency of the associated chloride channel (Guidotti et al., 1979; Nutt and Malizia, 2001). This mode of action assures a channel functioning very similar to physiological condition, while barbiturates, the former medications targeting $GABA_A$, act by a harsher mechanism, causing severe CNS depression, which can even lead to a fatal outcome.

Thus, BDZs represent ideal therapeutics for the treatment of a number of CNS disorders such as anxiety, insomnia, muscular spasticity, and epilepsy. Nevertheless, like other psychotropic agents, they are often abused or misused without a medical prescription.

Compounds based on a 1,4-BDZ scaffold were also identified as anticancer (Jones et al., 1990; Gill et al., 2014) and anti-HIV (Hsu et al., 1991) agents, all worth to be investigated in depth.

STRUCTURAL FEATURES OF BENZODIAZEPINES

Among the six different bicyclic isomers resulting from the fusion of a benzene and a diazepine ring, only derivatives of 1,4-BDZ system (Figure 2.1) possess sufficient requirements for clinical use. BDZs present on the market can be divided into two main subclasses: simple bicyclic derivatives, with International Non-proprietary Names generally characterized by a *-azepam* suffix, and tricyclic derivatives denoted with a *-zolam* suffix. A few examples of 1,5-BDZ derivatives, all sharing a *-azam* suffix, were shown to possess properties similar to 1,4-BDZ-related compounds (Pareek et al., 2013; Pasha and Jayashankara, 2006). The structures of all these molecules have in common an appended phenyl ring at position 5.

Due to their peculiar structure, these compounds display a lipophilic character and a weakly basic behavior at physiological *p*H, reflecting an ionization degree low enough to allow them to easily enter the CNS.

The aromatic six-membered fused ring and an electron-rich substituent at C2 represent prerequisites necessary for the specific binding to the $GABA_A$ receptor, while the appended aromatic substituent at C5 is required for gaining an agonistic activity. Each particular member of the 1,4-BDZ class is then characterized by substituents at different positions of the basic structure on both the aromatic rings and the diazepine moiety (Greenblatt et al., 1983). In particular, the majority of the commercially available drugs present an electron-withdrawing substituent at position 7, preferentially a halogen atom (chlorine or, less frequently, bromine) or a nitro group. The electron-rich substituent at C2 is generally an oxygen atom or a part of an extra heterocyclic ring condensed on side *a*. The appended phenyl ring is present as such or substituted at *ortho*-position with a chlorine or fluorine atom or replaced by a pyridine or a cyclohexene, while in some cases the benzo-fused ring is replaced

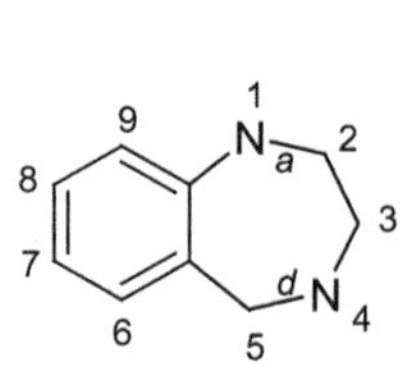

FIGURE 2.1 Skeleton of 1,4-BDZ.

by a bioisostere such as thiophene, pyrrole, pyrazole, or pyridine. A few examples of active compounds also contain a fused heterocyclic ring on side *d*. Some significant modifications are, however, compatible with activity especially in derivatives with prodrug characteristics. Benzophenone-based precursors able to cyclize *in vivo* have been proposed as BDZ prodrugs for particular applications (Yamamoto et al., 1984). Moreover, derivatives with a structure similar to that of 1,4-BDZ were shown to possess cardiovascular, analgesic, nootropic, or antidepressant effects, as well as other activities on CNS (Colussi et al., 2018; Filippakopoulos et al., 2012; Petty et al., 1995; Reddy and Patt, 1994).

DISCOVERY AND SYNTHESIS OF 1,4-BENZODIAZEPINES

Chlordiazepoxide (Librium®) was the first tranquilizer with a BDZ structure serendipitously discovered in 1957 by Dr Leo Sternbach at Roche. The research was initially devoted to the identification of tranquilizer agents with a structure other than that of BDZ, but the results were disappointing. During a successive reexamination, some interesting results were obtained for one of the previously synthesized compounds. A thorough structural analysis of the compound revealed its unexpected BDZ nature. In fact, the reaction of 6-chloro-2-(chloromethyl)-4-phenylquinazoline-3-oxide with methylamine did not result in the desired amine-quinazoline derivative (Figure 2.2, left), but generated chlordiazepoxide by an intramolecular rearrangement (Figure 2.2, right) (Sternbach, 1980).

Some synthetic routes have been then pursued for the specific preparation of BDZ derivatives (Nutley and Sternbach, 1968). Representatively, the general synthetic pathway successively developed by Sternbach leading to diazepam, probably the most studied BDZ, is reported in Figure 2.3. This general approach is based on the acylation of 5-chloro-2-(methylamino)benzophenone with chloroacetyl chloride to give an amide intermediate. The latter was in turn treated with hexamine to generate the corresponding amine-amide, which undergoes intramolecular cyclization leading to diazepam (Archer and Sternbach, 1968). Alternatively, the starting ketone was reacted with ethyl glycinate to directly obtain the desired final product, although in a lower yield. This scheme was then opportunely adapted for the preparation of numerous biologically active BDZ derivatives (Battle et al., 2018).

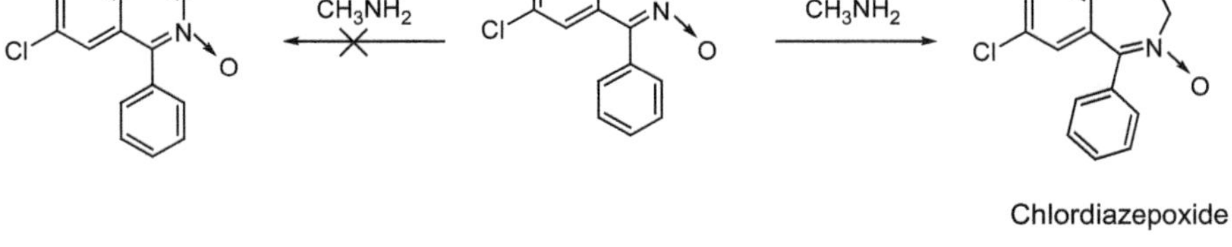

FIGURE 2.2 Attempted preparation of an amine-quinazoline (left); unexpected formation of chlordiazepoxide (right).

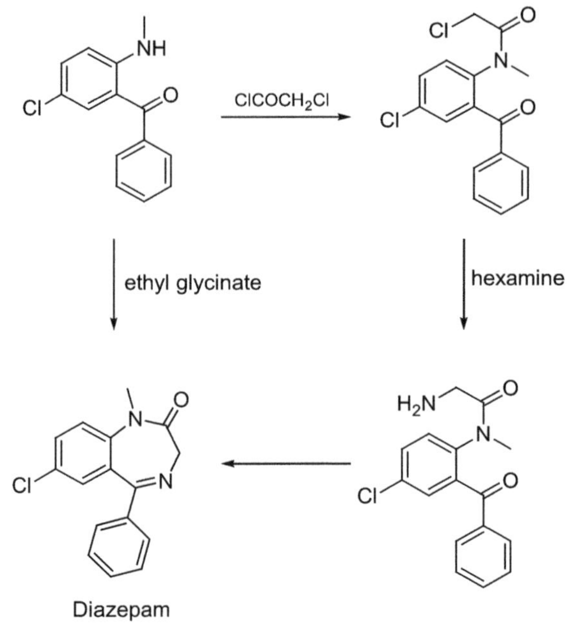

FIGURE 2.3 Synthesis of diazepam.

CLASSIFICATION OF BENZODIAZEPINES ACCORDING TO ELIMINATION HALF-LIFE

Chemical substituents on BDZ scaffold strongly affect potency and action onset and duration. This latter pharmacokinetic parameter has been adopted in order to get an overall BDZ classification (Greenblatt et al., 1981; Hagan, 1995; Ito et al., 1993; Mihic and Harris, 2015). Accordingly, based on their half-life, BDZs can be classified as follows:

- **Long-acting**: Elimination half-life >48 h, mainly used as anxiolytics (Figure 2.4)
- **Intermediate-acting**: Elimination half-life of 24–48 h, mainly used as anxiolytics and anticonvulsants (Figure 2.5)
- **Short-acting**: Elimination half-life <24 h, mainly used as hypnotics and anxiolytics (Figure 2.6)
- **Very short-acting**: Elimination half-life of 1–7 h, mainly used as hypnotics and in anesthesia procedures (Figure 2.7).

However, the duration of action is often dependent on the half-life of active intermediate metabolites rather than on the half-life of the parent drug. The most numerous group is represented by the long-acting agents that are usually transformed into still active metabolites. A further prolongation of action duration could be sometimes achieved by the administration of opportunely designed prodrugs.

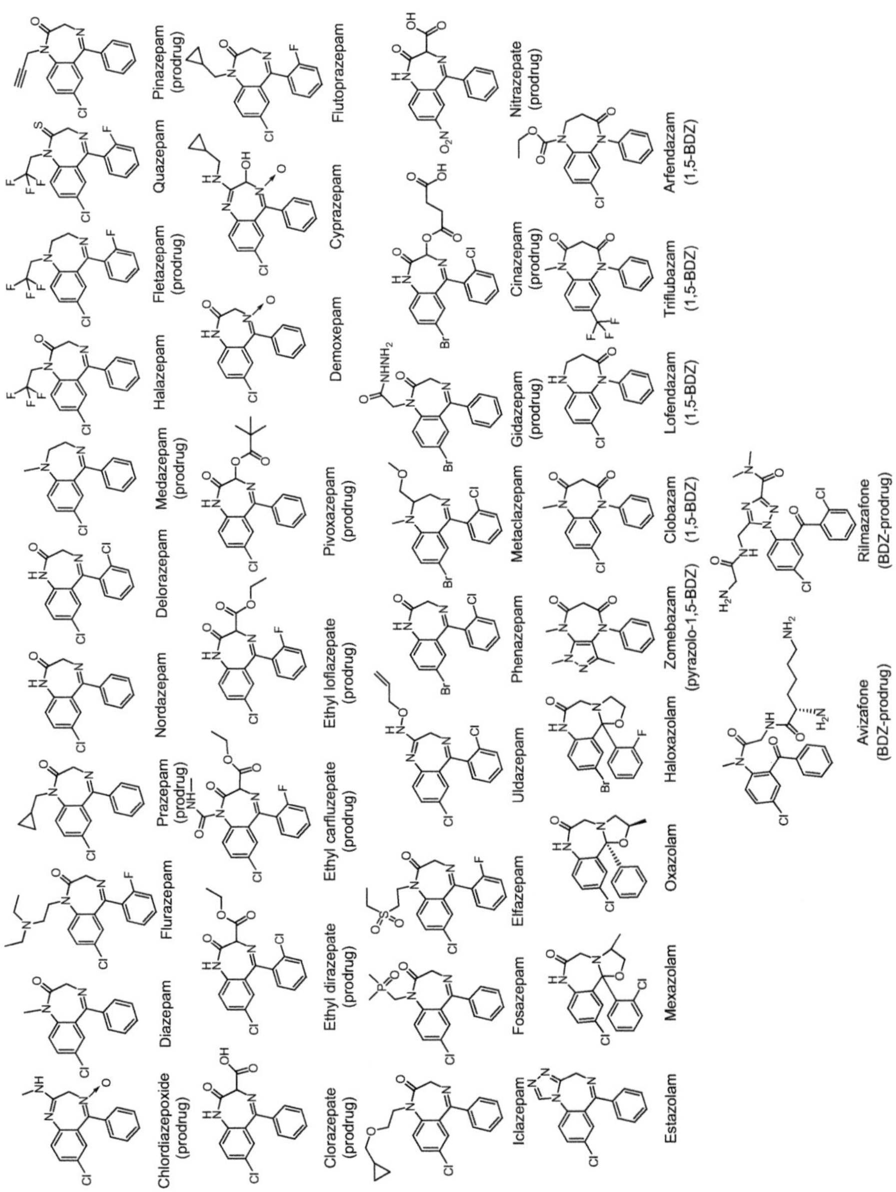

FIGURE 2.4 Chemical structures of long-acting BDZ.

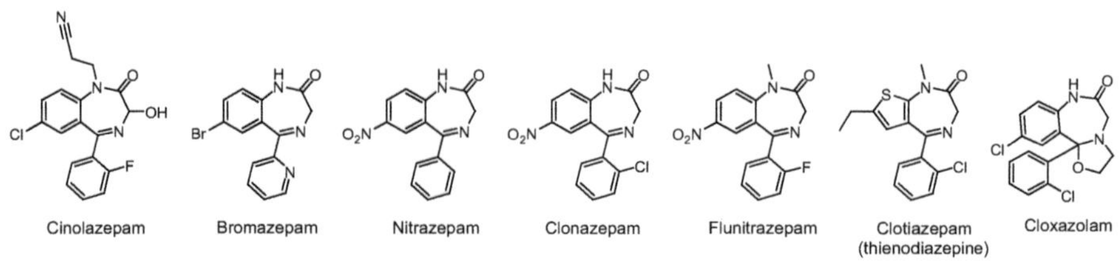

FIGURE 2.5 Chemical structures of intermediate-acting BDZ.

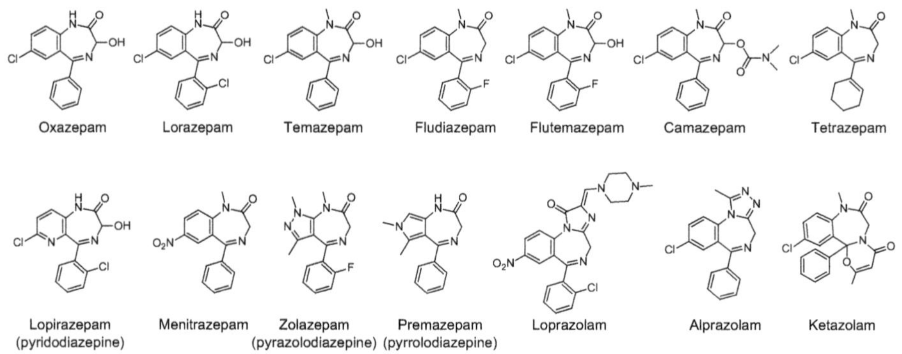

FIGURE 2.6 Chemical structures of short-acting BDZ.

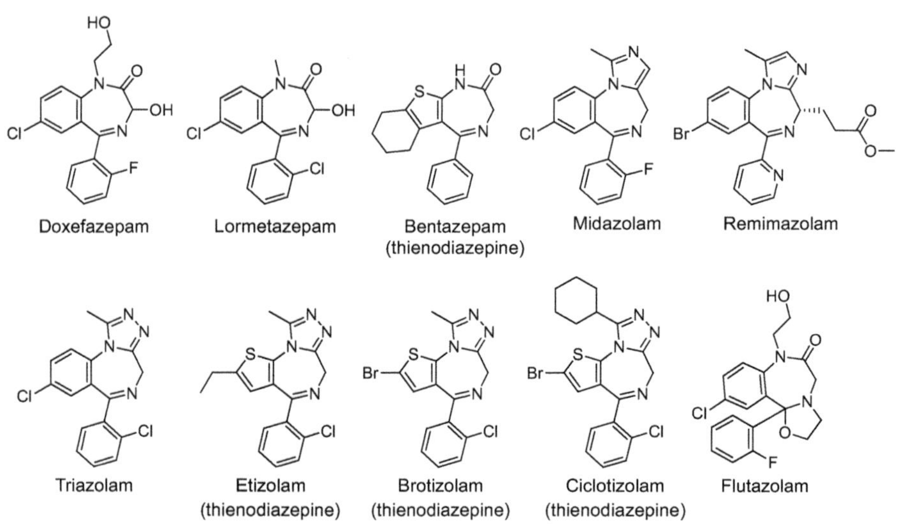

FIGURE 2.7 Chemical structures of very short-acting BDZ.

STRUCTURE–ACTIVITY RELATIONSHIPS (SAR)

The inspection of numerous BDZ structures has allowed to highlight a series of significant SAR (Sobanska et al., 2015; Valdes and Bayod, 2011; Young and Glennon, 1987), often confirmed by specific computational studies. With a few exceptions, the main SAR are as follows:

- The condensed benzo and 1,4-diazepine rings together with the appended aromatic ring are necessary for the activity;
- High receptor affinity is associated with the presence of an electron-withdrawing group, such as Cl, Br, or NO_2 at position 7, the former being the most frequent, while the nitro group confers hypnotic activity;
- Substitutions on other positions of the benzo-fused ring cause a decrease or even suppression of activity;
- Ortho position of the appended ring may be unsubstituted or bears a halogen atom (Cl or F), the presence of which usually increases activity, whereas substitutions at different positions are detrimental for activity;
- With a few exceptions, a carbonyl oxygen is present at position 2. This feature is obviously lacking in several tricyclic derivatives, where this electron-rich portion is replaced by atoms of a nitrogen-containing heterocycle (triazole or imidazole) fused at side *a*;
- The position 3 is generally unsubstituted or bears a hydroxyl group or a carboxyl group, either free or substituted. A few derivatives, with a prodrug characteristic, show a free or masked carboxyl group on this position.

BENZODIAZEPINES "DESIGNER DRUGS"

Similar to other psychoactive drugs, BDZ prescriptions are subjected to legal restrictions worldwide in order to avoid any misuse. Designer drugs (also termed as legal highs) are molecules designed to circumvent legal issues (Langston, 1987; Moosmann and Auwarter, 2018; Rozenek et al., 2019; Ziporyn, 1986). These compounds are conceived to display pharmacological effects similar to the parent drugs, eluding law restrictions and misleading standard drug tests (Wohlfarth and Weinmann, 2010). Among these, a number of new BDZs have been clandestinely synthesized and illegally marketed (Figure 2.8) (Zawilska and Wojcieszak, 2019).

IMIDAZO-BENZODIAZEPINES

A largely studied BDZ subclass is represented by the imidazo-BDZ obtained by the condensation on side *a* of the diazepine moiety with an imidazole ring, and lacking the phenyl substituent (Yang et al., 2009). This particular scaffold closely resembles that of classical BDZ drugs, but the absence of the appended ring confers several peculiar pharmacological properties, other than pure positive allosteric modulation of $GABA_A$ (Hunkeler et al., 1981). In particular, the most studied member of this class is represented by flumazenil (Haefely and Hunkeler, 1988). This compound acts as a competitive antagonist at the BDZ binding site on the $GABA_A$ receptor, so it

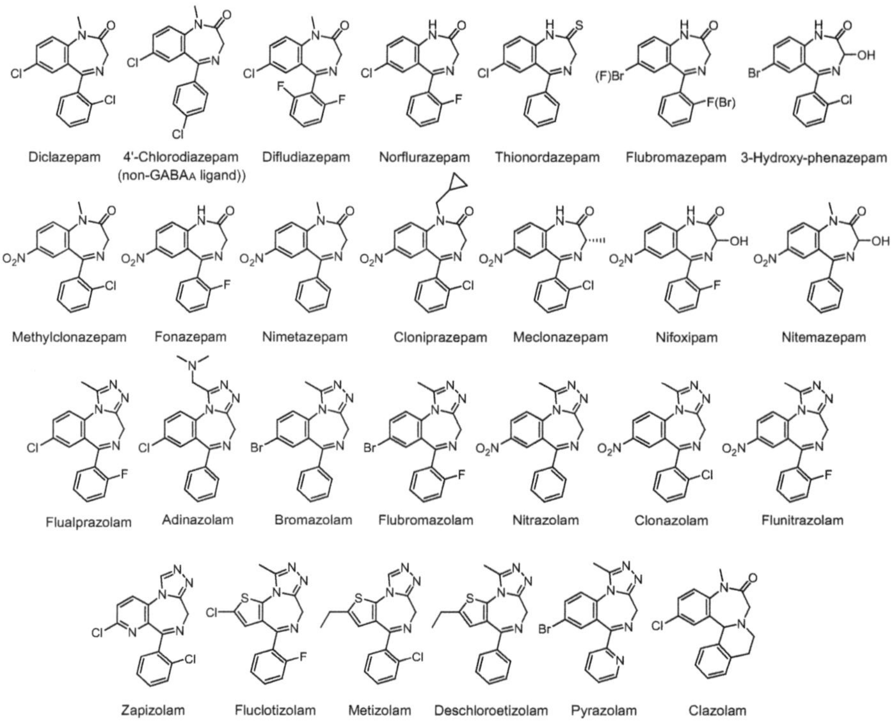

FIGURE 2.8 Chemical structures of BDZ "designer drugs."

is used for the treatment of BDZ overdose and for helping reverse anesthesia (Araki et al., 2005; Lee et al., 2018; Pani et al., 2015; Penninga et al., 2016). In fact, this agent binds with high affinity to the receptor producing negligible effects, thus preventing or reversing the effects of classical BDZ (Martinez-Cue et al., 2014; Pieri et al., 1981).

Several other related compounds, often denoted by a *-azenil* suffix, act as antagonists, partial or full agonists, or partial inverse agonists. As a unique exception, imidazenil, an imidazo-derivative bearing a 2'-bromophenyl at position 5, still acts as a partial agonist (Figure 2.9) (Serra et al., 1994).

NATURALLY OCCURRING BENZODIAZEPINES

The 1,4-BDZ system, even if embedded in more complex structures, has been found in several naturally occurring alkaloids. In particular, a series of polycyclic compounds showing interesting biological activity were assigned a BDZ-quinazoline or BDZ-oxepinopyrimidine scaffold. These compounds have been isolated from several genres of fungi such as *Penicillium* and *Aspergillus*. The most representative examples of these tetracyclic compounds are sclerotigenin and benzomalvins A–C isolated from *Penicillium* species (Joshi et al., 1999; Sun et al., 1994), asperlicins (Goetz et al., 1985; Liesch et al.,1985; Sun et al., 1994), and circumdatins A, B and L (Dai et al., 2001; Kshirsagar, 2015; Peng et al., 2013; Rahbaek et al., 1999), isolated from the fungus *Aspergillus ochraceus* (Figure 2.10). Some of these derivatives have been

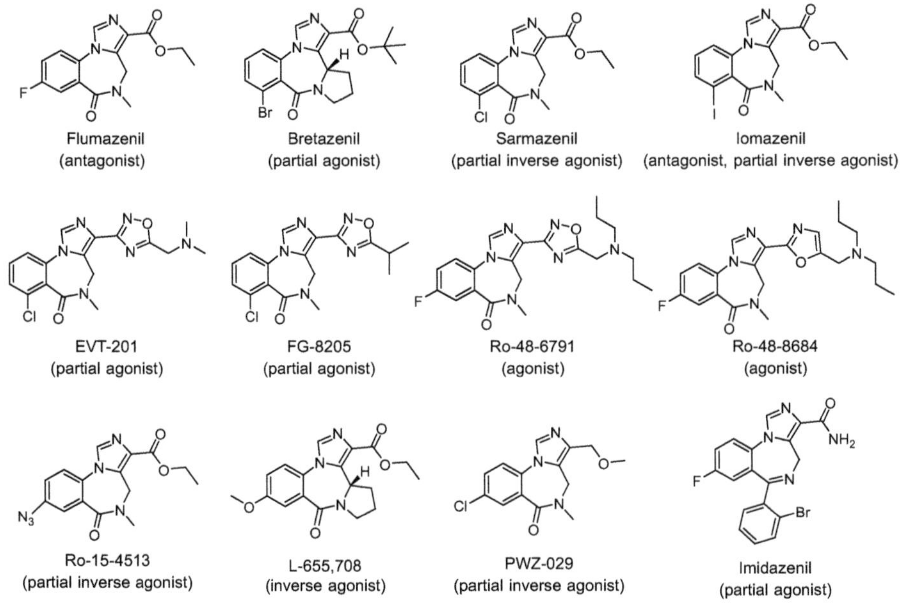

FIGURE 2.9 Chemical structures of imidazo-BDZ.

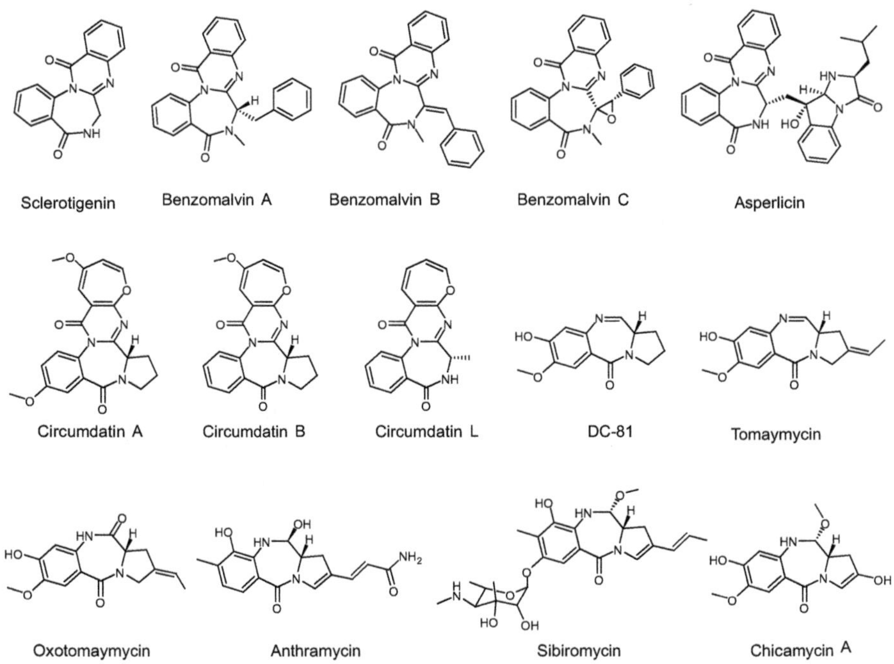

FIGURE 2.10 Chemical structures of naturally occurring BDZ.

proposed for the treatment of gastrointestinal disorders. Another group of naturally occurring BDZ-based substances is represented by the family of pyrrolo[2,1-*c*][1,4] BDZ antibiotics isolated from various species of *Streptomyces* and endowed with a broad-spectrum antitumor activity (Annor-Gyamfi et al., 2018). The most studied derivatives of this series, DC-81, anthramycin, and sibiromycin, although cardiotoxic, were adopted as lead compounds for the development of potent anticancer agents (Hu et al., 2003; Hurley et al., 1979).

Z-DRUGS

A group of anxiolytic agents structurally unrelated to BDZ and known as Z-drugs entered the market around the end of the 1980s. Initially, only three molecules were commercialized, namely, zolpidem (with an imidazopyridine scaffold), zopiclone (a cyclopyrrolone), and zaleplon (a pyrazolopyrimidine), as hypnotics and tranquillizers with an initial toxicological profile even safer than that of BDZ (Figure 2.11) (Roehrs and Roth, 2010; Siriwardena et al., 2008). These alternative sedative–hypnotic drugs were endowed with a peculiar pharmacokinetic profile, which includes a rapid onset of action and a short half-life, thus providing a low residual effect as compared with BDZ. Despite the different structure, their mechanism of action closely resembles that of BDZ showing a higher selectivity for BZ1 receptor (α1-subunit-containing $GABA_A$ subclass). On the other hand, these agents also exert unwanted effects, such as amnesia and visual hallucinations. In long-term studies, the use of these drugs produced, however, the same drawbacks associated with BDZ. Moreover, oral administration of these drugs in conjunction with alcohol or other psychotropic agents could be the cause of severe problems (Gunja, 2013). Following the market entry of the first three Z-drugs, many other similar agents were produced and clinically tested, but most of them did not successfully pass clinical trials or were discontinued from the market (Brandt and Leong, 2017; Verthein et al., 2019).

CONCLUSIONS

Since their first identification, BDZs almost completely replaced the former remedies for the treatment of insomnia and anxiety due to their high efficacy and favorable toxicological profile. These properties, together with the optimization of very simple synthetic methods, gave rise to a huge armamentarium of compounds, which allowed in-depth investigations concerning their mode of action and SAR. Nevertheless, the research and development of alternative synthetic methods (Tao et al., 2020), as well as the identification of novel and safe pharmacologically active analogues, are still ongoing (Di Capua et al., 2020). However, despite the approval of new related agents belonging to other chemical classes, BDZs still remain among the most prescribed drugs worldwide.

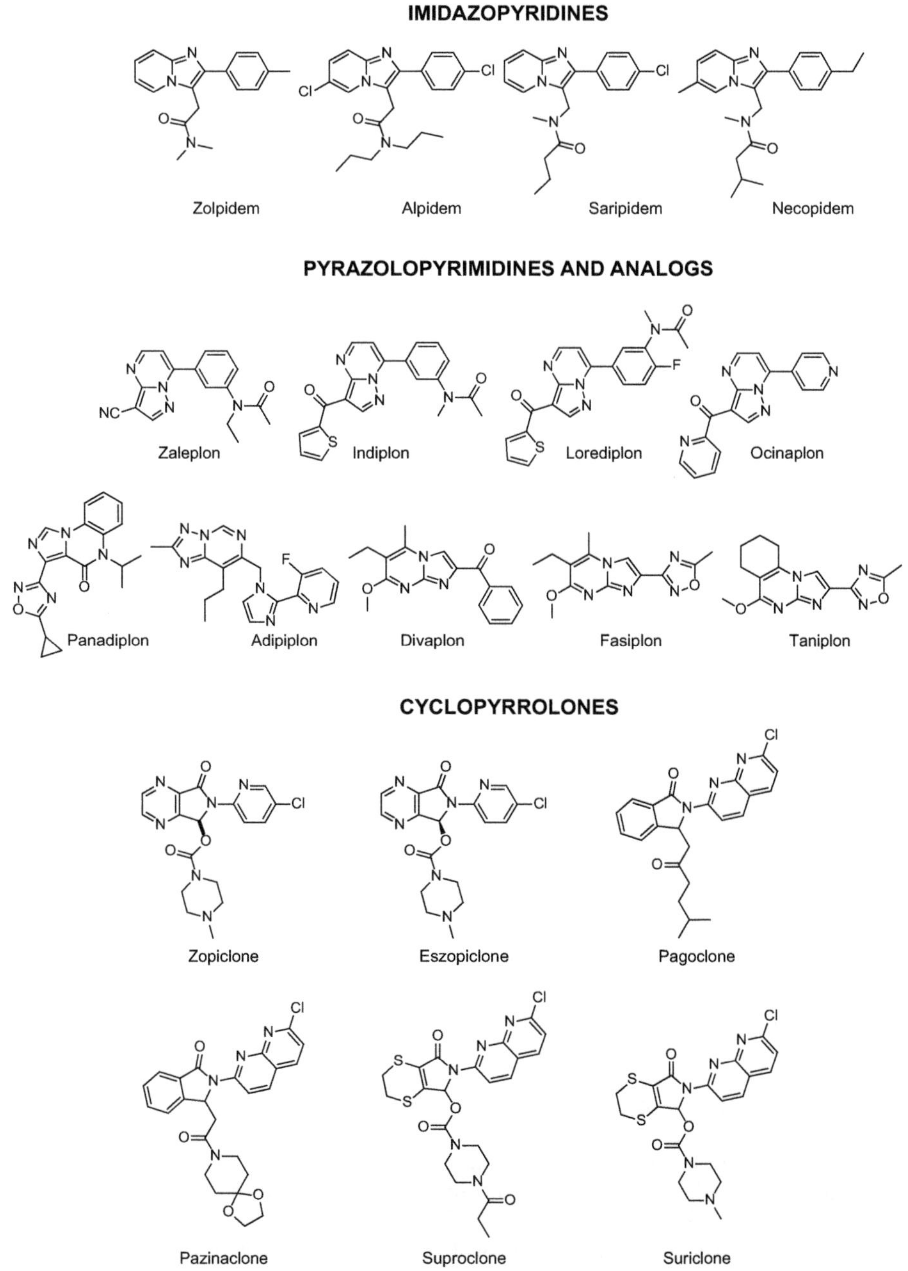

FIGURE 2.11 Chemical structures of Z-drugs or non-BDZ.

REFERENCES

Annor-Gyamfi JK, Jarrett JM, Osazee JO, Bialonska D, Whitted C, Palau VE, Shilabin AG. Synthesis and biological activity of fused tetracyclic Pyrrolo[2,1-*c*][1,4]benzodiazepines. *Heliyon* 2018; 4(2): e00539.

Araki H, Fujiwara Y, Shimada Y. Effect of flumazenil on recovery from sevoflurane anesthesia in children premedicated with oral midazolam before undergoing herniorrhaphy with or without caudal analgesia. *J Anesth* 2005; 19(3): 204–207.

Archer GA, Sternbach LH. Chemistry of benzodiazepines. *Chem Rev* 1968; 68(6): 747–784.

Battle E, Lizano E, Vinas M, Pujol MD. 1,4-Benzodiazepines and new derivatives: description, analysis, and organic synthesis. *Med Chem Chap* 2018; 5: 63, 10.5772/intechopen.79879.

Brandt J, Leong C. Benzodiazepines and Z-drugs: an updated review of major adverse outcomes reported on in epidemiologic research. *Drugs R D* 2017; 17(4): 493–507.

Colussi G, Catena C, Darsie D, Sechi LA. Benzodiazepines: an old class of new antihypertensive drugs? *Am J Hypertens* 2018; 31(4): 402–404.

Dai J, Carte BK, Sidebottom PJ, Sek Yew A L, Ng S, Huang Y, Butler MS. Circumdatin G, a new alkaloid from the fungus *Aspergillus ochraceus*. *J Nat Prod* 2001; 64(1): 125–126.

Di Capua A, Reale A, Paolino M, Chemi G, Brogi S, Cappelli, A, Giorgi G, Grande F, Di Cesare Mannelli L, Ghelardini C, Matucci R, Garofalo A, Anzini M. Design, synthesis and biological evaluation of 7-substituted 4-phenyl-6H-imidazo[1,5-*a*]thieno[3,2-*f*][1,4] diazepines as safe anxiolytic agents. *Eur J Med Chem* 2020; 200: 112405.

Filippakopoulos P, Picaud S, Fedorov O, Keller M, Wrobel M, Morgenstern O, Bracher F, Knapp S. Benzodiazepines and benzotriazepines as protein interaction inhibitors targeting bromodomains of the BET family. *Bioorg Med Chem* 2012; 20(6): 1878–1886.

Gill RK, Kaushik SO, Chugh J, Bansal S, Shah A, Bariwal J. Recent development in [1,4] benzodiazepines as potent anticancer agents: a review. *Mini Rev Med Chem* 2014; 14 (3): 229–256.

Goetz MA, Lopez M, Monaghan RL, Chang RS, Lotti VJ, Chen TB. Asperlicin, a novel nonpeptidal cholecystokinin antagonist from *Aspergillus alliaceus*. Fermentation, isolation and biological properties. *J Antibiot (Tokyo)* 1985; 38(12): 1633–1637.

Greenblatt DJ, Shader RI, Abernethy DR. Drug therapy. Current status of benzodiazepines. *N Engl J Med* 1983; 309(6): 354–358.

Greenblatt DJ, Shader RI, Divoll M, Harmatz JS. Benzodiazepines: a summary of pharmacokinetic properties. *Br J Clin Pharmacol* 1981; 11 Suppl 1: 11S–16S.

Guidotti A, Baraldi M, Schwartz JP, Costa E. Molecular mechanisms regulating the interactions between the benzodazepines and GABA receptors in the central nervous system. *Pharmacol Biochem Behav* 1979; 10(5): 803–807.

Gunja N. The clinical and forensic toxicology of Z-drugs. *J Med Toxicol* 2013; 9(2): 155–162.

Haefely W, Hunkeler W. The story of flumazenil. *Eur J Anaesthesiol Suppl* 1988; 2: 3–13.

Hagan RL. Clarification of benzodiazepine structural classes. *J Anal Toxicol* 1995; 19(1): 58–59.

Hsu MC, Schutt AD, Holly M, Slice LW, Sherman MI, Richman DD, Potash MJ, Volsky DJ. Inhibition of HIV replication in acute and chronic infections in vitro by a Tat antagonist. *Science* 1991; 254(5039): 1799–1802.

Hu WP, Yu HS, Chen YC, Wang JJ. Biological evaluation of an antibiotic DC-81-indole conjugate agent in human melanoma cell lines. *Kaohsiung J Med Sci* 2003; 19(1): 6–12.

Hunkeler W, Mohler H, Pieri L, Polc P, Bonetti EP, Cumin R, Schaffner R, Haefely W. Selective antagonists of benzodiazepines. *Nature* 1981; 290(5806): 514–516.

Hurley LH, Allen CS, Feola JM, Lubawy WC. *In vitro* and *in vivo* stability of anthramycin-DNA conjugate and its potential application as an anthramycin prodrug. *Cancer Res* 1979; 39(8): 3134–3140.

Ito K, Yamada Y, Nakamura K, Sawada Y, Iga T. Classification of benzodiazepine hypnotics in humans based on receptor occupancy theory. *J Pharmacokinet Biopharm* 1993; 21(1): 31–41.

Jones G B, Davey CL, Jenkins TC, Kamal A, Kneale GG, Neidle S, Webster GD, Thurston DE. The non-covalent interaction of pyrrolo[2, 1-*c*][1,4]benzodiazepine-5,11-diones with DNA. *Anticancer Drug Des* 1990; 5(3): 249–264.

Joshi BK, Gloer JB, Wicklow DT, Dowd PF. Sclerotigenin: a new antiinsectan benzodiazepine from the sclerotia of penicillium sclerotigenum. *J Nat Prod* 1999; 62(4): 650–652.

Kshirsagar UA. Recent developments in the chemistry of quinazolinone alkaloids. *Org Biomol Chem* 2015; 13(36): 9336–9352.

Kyburz E. Serendipity in drug discovery. The case of BZR ligands. *Farmaco* 1989; 44(4): 345–382.

Langston JW. Neurology: neurologic consequences of designer drugs. *West J Med* 1987; 146(3): 351.

Lee SP, Sung IK, Kim JH, Lee SY, Park HS, Shim CS. Efficacy and safety of flumazenil injection for the reversal of midazolam sedation after elective outpatient endoscopy. *J Dig Dis* 2018; 19(2): 93–101.

Liesch JM, Hensens OD, Springer JP, Chang RS, Lotti VJ. Asperlicin, a novel non-peptidal cholecystokinin antagonist from *Aspergillus alliaceus*. Structure elucidation. *J Antibiot (Tokyo)* 1985; 38(12): 1638–1641.

Maramai S, Benchekroun M, Ward SE, Atack JR. Subtype selective gamma-aminobutyric acid type A receptor ($GABA_AR$) modulators acting at the benzodiazepine binding site: an update. *J Med Chem* 2020; 63(7): 3425–3446.

Martinez-Cue C, Delatour B, Potier MC. Treating enhanced GABAergic inhibition in Down syndrome: use of GABA alpha5-selective inverse agonists. *Neurosci Biobehav Rev* 2014; 46 Pt 2: 218–227.

McKernan RM, Rosahl TW, Reynolds DS, Sur C, Wafford KA, Atack JR, Farrar S, Myers J, Cook G, Ferris P, Garrett L, Bristow L, Marshall G, Macaulay A, Brown N, Howell O, Moore KW, Carling RW, Street L J, Castro JL, Ragan CI, Dawson GR, Whiting PJ. Sedative but not anxiolytic properties of benzodiazepines are mediated by the GABA(A) receptor alpha1 subtype. *Nat Neurosci* 2000; 3(6): 587–592.

Mihic SJ, Harris AR. Hypnotic and Sedatives Goodman & Gilman's The Pharmacological Basis of Therapeutics, *Goodman & Gilman's The Pharmacological Basis of Therapeutics* 2015; Chapter 17, McGraw-Hill Education/Medical.

Moosmann B, Auwarter V. Designer benzodiazepines: another class of new psychoactive substances. *Handb Exp Pharmacol* 2018; 252: 383–410.

Nutley ER, Sternbach LH. 5-aryl-3*H*-1,4-benzodiazepin-2(1*H*)-ONES 1968; *US3371085*.

Nutt DJ, Malizia AL. New insights into the role of the GABA(A)-benzodiazepine receptor in psychiatric disorder. *Br J Psychiatry* 2001; 179: 390–396.

Pani N, Dongare PA, Mishra RK. Reversal agents in anaesthesia and critical care. *Indian J Anaesth* 2015; 59(10): 664–669.

Pareek A, Kumar N, Agarwal A, Sharma P, Dharma K. 1,5-Benzodiazepines: overview of properties and synthetic aspects. *Res J Chem Sci* 2013; 3(7): 90–103.

Pasha MA, Jayashankara V P. An expeditious synthesis of 1,5-benzodiazepine derivatives catalyzed by CdCl2. *Indian J Chem B* 2006; 45(12): 2716–2719.

Peng J, Zhang XY, Tu Z C, Xu XY, Qi SH. Alkaloids from the deep-sea-derived fungus *Aspergillus westerdijkiae* DFFSCS013. *J Nat Prod* 2013; 76(5): 983–987.

Penninga EI, Graudal N, Ladekarl MB, Jurgens G. Adverse events associated with flumazenil treatment for the management of suspected benzodiazepine intoxication—a systematic review with meta-analyses of randomised trials. *Basic Clin Pharmacol Toxicol* 2016; 118(1): 37–44.

Petty F, Trivedi MH, Fulton M, Rush AJ. Benzodiazepines as antidepressants: does GABA play a role in depression? *Biol Psychiatry* 1995; 38(9): 578–591.

Pieri L, Schaffner R, Scherschlicht R, Polc P, Sepinwall J, Davidson A, Mohler H, Cumin R, Da Prada M, Burkard WP, Keller HH, Muller RK, Gerold M, Pieri M, Cook L, Haefely W. Pharmacology of midazolam. *Arzneimittelforschung* 1981; 31(12a): 2180–2201.

Rahbaek L, Breinholt J, Frisvad JC, Christophersen C. Circumdatin A, B, and C: three new benzodiazepine alkaloids isolated from a culture of the fungus *Aspergillus ochraceus. J Org Chem* 1999; 64(5): 1689–1692.

Reddy S, Patt RB. The benzodiazepines as adjuvant analgesics. *J Pain Symptom Manage* 1994; 9(8): 510–514.

Roehrs T, Roth T. Drug-related sleep stage changes: functional significance and clinical relevance. *Sleep Med Clin* 2010; 5(4): 559–570.

Rozenek EB, Wilczynska K, Gorska M, Waszkiewicz N. Designer drugs - still a threat? *Przegl Epidemiol* 2019; 73(3): 337–347.

Serra M, Ghiani CA, Motzo C, Cuccheddu T, Floris S, Giusti P, Biggio G. Imidazenil, a new partial agonist of benzodiazepine receptors, reverses the inhibitory action of isoniazid and stress on gamma-aminobutyric acid A receptor function. *J Pharmacol Exp Ther* 1994; 269(1): 32–38.

Siriwardena AN, Qureshi MZ, Dyas JV, Middleton H, Orner R. Magic bullets for insomnia? Patients' use and experiences of newer (Z drugs) versus older (benzodiazepine) hypnotics for sleep problems in primary care. *Br J Gen Pract* 2008; 58(551): 417–422.

Sobanska AW, Zydek G, Wlodno P, Brzezinska E. Comparative (Q)SAR analysis of benzodiazepine derivatives with different biological activity. *Eur J Med Chem* 2015; 89: 147–155.

Sternbach L. The Benzodiazepine Story, *Benzodiazepines Today and Tomorrow*, 1980; 5–17, MTP Press.

Sun HH, Barrow CJ, Sedlock DM, Gillum AM, Cooper R. Benzomalvins, new substance P inhibitors from a *Penicillium* sp. *J Antibiot (Tokyo)* 1994; 47(5): 515–522.

Tao S, Bu Q, Shi Q, Wei D, Dai B, Liu N. Synthesis of benzodiazepines through ring opening/ring closure of benzimidazole salts. *Chemistry* 2020; 26(15): 3252–3258.

Valdes C, Bayod M. The Chemistry of Benzodiazepines, *Modern Heterocyclic Chemistry*, 2011; First Edition. Wiley-VCH Verlag GmbH & Co. KGaA.

Verthein U, Buth S, Holzbach R, Neumann-Runde E, Martens MS. Benzodiazepines and Z-drugs—analyses of ambulatory prescriptions from 2006 to 2015. *Psychiatr Prax* 2019; 46(7): 399–405.

Wohlfarth A, Weinmann W. Bioanalysis of new designer drugs. *Bioanalysis* 2010; 2(5): 965–979.

Yamamoto K, Hirose K, Matsushita A, Yoshimura K, Sawada T, Eigyo M, Jyoyama H, Fujita A, Matsubara K, Tsukinoki Y, Hatakeyama H, Shiomi T, Ibii N, Inoue Y, Hirono S, Koshida H. Pharmacological studies of a new sleep-inducer, 1*H*-1,2,4-triazolyl benzophenone derivatives (450191-S) (I). Behavioral analysis. *Nihon Yakurigaku Zasshi* 1984; 84(1): 109–154.

Yang J, Teng Y, Ara S, Rallapalli S, Cook JM. An improved process for the synthesis of 4*H*-imidazo[1,5-*a*][1,4]benzodiazepines. *Synthesis (Stuttg)* 2009; 40(32): nihpa145687.

Young R, Glennon RA. Stimulus properties of benzodiazepines: correlations with binding affinities, therapeutic potency, and structure activity relationships (SAR). *Psychopharmacology (Berl)* 1987; 93(4): 529–533.

Zawilska JB, Wojcieszak J. An expanding world of new psychoactive substances-designer benzodiazepines. *Neurotoxicology* 2019; 73: 8–16.

Ziporyn T. A growing industry and menace: makeshift laboratory's designer drugs. *J Am Med Assoc* 1986; 256(22): 3061–3063.

3 Central Benzodiazepine Receptors: Structure and Function

Michael H. Ossipov
University of Arizona College of Medicine

CONTENTS

In the early 1930s, a Polish postgraduate chemist, Leo H. Sternbach, was interested in developing synthetic dyes and synthesized a number of heptoxdiazine compounds. Two decades later, he was working for Hoffman-La Roche in NJ, and inspired by the commercialization of the antipsychotic chlorpromazine, he began to experiment by adding side chains to his ersatz research interest—the heptoxdiazines. However, after synthesizing about 40 such compounds that were pharmacologically inert, one, Ro 5-0690, was stored on a shelf and collected dust until it was rediscovered sometime later during a laboratory cleanup. It was submitted for testing and was discovered to be superior to meprobamate as a muscle relaxant and anxiolytic [1–3]. It was also found to have similar sedative properties to chlorpromazine, but without "significant" adverse effects [2]. In an effort to discover why this compound had these properties when so many previous analogs failed, it was discovered that the use of methylamine in its synthesis altered the reaction compared to earlier ones, involving transposition reaction with ring enlargement, and formed 1,4-benzodiazepine, which entered clinical use as chlordiazepoxide (Librium®) when the US FDA approved it in 1960 for the relief of anxiety. This serendipitous discovery soon followed by the development of diazepam (Valium®) was approved in 1962 for the management of anxiety, skeletal muscle spasm, acute alcohol withdrawal, and as an adjunct in convulsive disorders. These developments were soon followed by the analogs nitrazepam, flunitrazepam, and clonazepam. Competing pharmaceutical companies soon joined in developing rival compounds, leading to the introduction of a new class of compounds—the benzodiazepines [4]. Work in this field has been quite prolific, with at least 17 different benzodiazepines currently in use. The variations in potency, time to onset, and duration of action provide for their use for several indications, including anxiety, insomnia, muscle relaxation, spasticity due to central nervous system (CNS) pathology, and epilepsy, as well as for intraoperative uses due to amnesic and antianxiety properties [5].

The introduction of chlordiazepoxide and diazepam to clinical practice was met with considerable enthusiasm because these drugs were viewed as replacements for barbiturates, which had problems with addiction liability, were addictive, required increasing dosages over time, and carried a risk of overdose and interactions with other medications [6, 7]. In contrast, the benzodiazepines were perceived as safe with no addiction liability. It was assumed that the low doses, and the fact that patients were not consistently asking for ever-increasing doses, were signs that these drugs had low addiction liability [4].

A key advantage of benzodiazepines over the barbiturates is that benzodiazepines cause significantly less respiratory depression, and when they are taken alone, overdose with benzodiazepines alone is rarely lethal [8]. In part because of their favorable safety profile and the perception that they had little abuse and dependence liability, the use of benzodiazepines increased rapidly throughout the 1970s and benzodiazepines were often among the most commonly prescribed drugs [9]. In 2009, there were over 100 million prescriptions written for benzodiazepines in the United States alone, with alprazolam, clonazepam, diazepam, and lorazepam being the four most widely prescribed [10].

There were some indications early on that benzodiazepines could be associated with some psychiatric adverse events early on. For example, there was awareness that chlordiazepoxide used in a US prison population was associated with dependence, but this was written off as a consequence of the higher doses used, and was not considered to be of concern with the lower doses used in everyday clinical practice [4]. Nevertheless, as the use of benzodiazepines increased over time, there came an increasing awareness that they may not be as free from negative consequences as initially believed, and as early as the 1980s, evidence of potential abuse and dependence was accumulating [9]. Although benzodiazepines have a low risk of toxicity when they are used alone, they act synergistically with opiates and ethanol, resulting in enhanced depression of the CNS and increased potential for fatal overdoses. Most instances of benzodiazepine overdose that are fatal include the concomitant use of alcohol or opiates.

Two notable cases from the 1980s serve to illustrate this dichotomy in benzodiazepine toxicity. In 1987, Admiral Robert McFarlane was caught up in a political scandal, and in desperation, he ingested "25 to 30" tablets of diazepam, and yet he was still conscious when the ambulance appeared, and recovered after a brief hospital stay [11]. This is consistent with the findings of a case report of two patients who ingested 500 and 2,000 mg of diazepam in suicide attempts, and were in a moderately deep coma, but recovered and were discharged within 48 h [12]. In contrast, the combination of benzodiazepines with other drugs, including ethanol, can be quite deadly. The second example from the 1980s is the tragic case of Karen Ann Quinlan, a young lady who, after ingesting benzodiazepines with alcohol, fell into a coma and remained in a persistent vegetative state for nearly 10 years before life-sustaining measures were terminated.

Tolerance to the different effects of benzodiazepines develops at variable rates, with tolerance to hypnotic effects occurring more rapidly than tolerance to antianxiety effects, and benzodiazepines may lose their efficacy after continued use over 4–6 months [8]. The use of benzodiazepines over a long period of time also results in impairment in memory and difficulty in coordinating movements with high risk of falls [10]. The most common withdrawal symptom from doses within the therapeutic range is anxiety, and autonomic instability, insomnia, and sensory hypersensitivity can also occur [8]. Abrupt withdrawal from benzodiazepines after long-term use (i.e., ≥6 months) can lead to life-threatening seizures, delirium, and death [10]. In this chapter, we review the properties of the central benzodiazepine receptor that underlie the range in activities of benzodiazepines.

BENZODIAZEPINE RECEPTOR—STRUCTURE

Benzodiazepines do not act at an independent receptor that directly affects the activity of the neuron they are attached to, but rather, it is an allosteric modulatory binding site on the $GABA_A$ (γ-aminobenzoic acid) receptor that enhances the activity of the endogenous ligand, GABA, which is the most abundant inhibitory neurotransmitter in the CNS. The $GABA_A$ receptor is a ligand-gated ion channel. When GABA binds to its recognition site on the channel, a conformational change in the chloride (Cl^-) ion channel occurs that opens the pore and allows the flow of Cl^- through it, resulting in inhibitory postsynaptic potentials and hyperpolarization of the neuron. Benzodiazepines, by binding to an allosteric modulatory site, increase the frequency of channel opening induced by GABA [7, 13–15].

The $GABA_A$ receptor is truly a fascinating structure. It consists of five protein subunits that are arranged in a circular formation, thus forming a central pore (Figure 3.1). The subunits themselves occur in several types, and these subunits each have several subtypes, explained in more detail below. What is remarkable is that the combination of these subunits results in $GABA_A$ receptors with different properties and differing sensitivities to allosteric ligands such as benzodiazepines, barbiturates, and alcohol. It is as if this one receptor is designed to have different properties, depending on its makeup, and consequently, different effects within the CNS, depending on the distribution of these different arrangements.

Each of the protein subunits that make up a $GABA_A$ receptor has a molecular weight of about 50 kDa and consists of approximately 450 amino acid residues. Approximately one-half of the protein chain forms the hydrophilic extracellular N-terminal, which may contain a recognition site for GABA. The $GABA_A$ receptor belongs to the Cys loop-type ligand-gated ion channel superfamily, which shares a highly conserved region of 13 amino acids between 2 cysteine (Cys) residues connected by a disulfide bond N-terminal, forming the loop. An intracellular loop between the 3rd and 4th transmembrane domains participates in the phosphorylation-mediated modulation of the receptor.

Although a $GABA_A$ receptor consists of 5 subunits, there are 19 different subunits available with which to construct the receptor, identified as $\alpha_{(1-6)}$, $\beta_{(1-3)}$, $\gamma_{(1-3)}$, δ, ε,

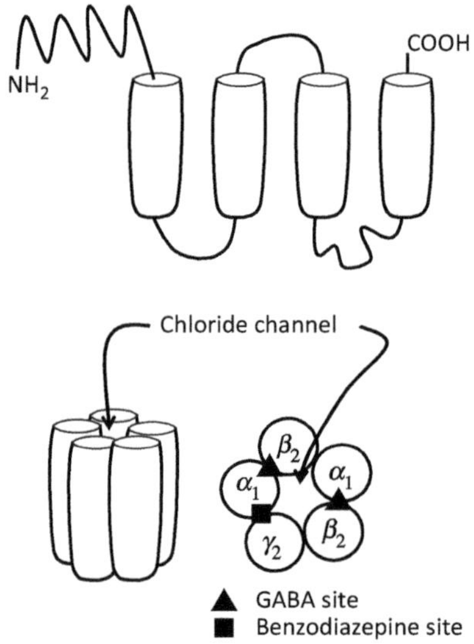

FIGURE 3.1 Schematic representation of the $GABA_A$ receptor is shown. Each receptor is composed of five subunits, and each subunit consists of four helical transmembrane domains and an intracellular loop. The five subunits of the $GABA_A$ receptor are arranged in a circular pattern to form a central chloride channel. The GABA recognition sites occur where an α- and a β-subunit meet, and the benzodiazepine site occurs at the interface between the α- and γ-subunits.

θ, π, and $\rho_{(1-3)}$, thus creating the potential for the existence of a myriad of receptor isoforms, each with different pharmacologic profiles. It is still unclear how many of the possible isotypes of the $GABA_A$ receptor actually occur in nature. Receptors with different subunit compositions are differentially expressed in the CNS, effectively meaning that different brain regions can be differentially sensitive to GABA and to the various modulators of GABA activity, such as ethanol, benzodiazepines, barbiturates, or the sedative steroids, due to these differences. For example, some subunits are broadly expressed throughout the CNS, others have a somewhat restricted distribution, and the α_6-subunit is expressed only on cerebellar granule neurons. Most $GABA_A$ receptors consist of 2 α-subunits, 2 β-subunits, and 1 γ-subunit [16].

In an electrophysiologic study, HEK cells were transfected with cDNA fragments coding for different isoforms of the $GABA_A$ receptor. It was found that GABA had the lowest potency at receptors expressing the α_2- or α_3-subunits, and the highest potency for those containing the α_6-subunit. Interestingly, the differences in affinity of GABA among these receptor isoforms ranged up to 175-fold. The affinities of GABA for different isoforms of the $GABA_A$ receptors are summarized in Table 3.1 [16].

The $GABA_A$ receptors occur at postsynaptic sites, where the release of GABA from a nerve terminal results in opening of chloride channels and hyperpolarization of the postsynaptic neuron. This phasic inhibition event occurs rapidly, within milliseconds. There are also extrasynaptic $GABA_A$ receptors that respond to ambient levels of GABA and maintain a persistent inhibitory tone [17]. $GABA_A$ receptors containing the δ-subunit are only found at extrasynaptic sites. These extrasynaptic

TABLE 3.1
GABA Potency and Maximal Currents at Different Subunit Compositions

Subunit Composition	EC_{50} (μM)	Maximum Current (pA)	CNS Distribution
α1β3γ2S	2.1	3367 ± 662	Widespread
α2β3γ2S	13.4	3056 ± 435	Widespread
α3β3γ2S	12.5	3776 ± 305	Thalamic and hypothalamic nuclei, locus coeruleus, dentate granule cells
α4β3γ2S	2.1	2574 ± 292	Thalamus
α5β3γ2S	1.4	2446 ± 445	Hippocampus
α6β3γ2S	0.17	2446 ± 445	Cerebellum and cochlear granule cells
α1β1γ2S	10.9	3575 ± 799	
α1β2γ2S	6.6	2230 ± 193	Widespread
Subunits listed below are found only at extrasynaptic sites			
α4β3	0.97	328 ± 67	Thalamus
α4β3δ	1.7	1224 ± 264	Thalamus
a6β3	0.076	490 ± 125	Cerebellar granule cells
α6β3δ	0.17	706 ± 148	Cerebellar granule cells
α1β2	1.7	1863 ± 333	Widespread
a3β3	4.5	3924 ± 288	Thalamic and hypothalamic nuclei, locus coeruleus
α1β2δ	3.7	398 ± 147	Hippocampus
α4β2δ	0.91	1544 ± 263	Hippocampus
a3β3θ	3.4	1680 ± 508	Hypothalamic nuclei, locus coeruleus
a3β3ε	0.86	811 ± 300	Hypothalamic nuclei, locus coeruleus

$GABA_A$ receptor subtypes are highly sensitive to neurosteroids, and dysfunction of these receptors is likely to be linked to conditions with reduced seizure thresholds and with certain anxiety disorders [17–19].

Endogenous modulators of the $GABA_A$ receptor are neurosteroids and the endocannabinoid 2-arachidonoyl glycerol (2-AG). There are also numerous exogenous small molecules that modulate the $GABA_A$ receptor. It is believed that the diversity of compounds that can interact with the $GABA_A$ receptor is due to the numerous pockets, or cavities, that form in the transmembrane portions of the $GABA_A$ receptor subunits [19, 20]. Computational modeling indicates that different subunits of the $GABA_A$ receptors form numerous cavities with different conformations such that they act as binding sites for different classes of drugs that can alter the functioning of the $GABA_A$ receptor. Several different sites on the $GABA_A$ receptor have been identified that have selectivity for the endogenous substrates 2-AG and neurosteroids, as well as for benzodiazepines, ethyl alcohol, barbiturates, and general anesthetics. Occupation of one of these cavities by a substrate causes changes in receptor conformation, and can enhance binding of GABA to the receptor and/or alter the characteristics of the channel to remain open longer or more frequently. Consequently, benzodiazepines and other substances, such as alcohol, are in a position to synergize each other's effects at the GABA receptor.

Benzodiazepine Modulation of the $GABA_A$ Receptor

Although there are many possible configurations, most $GABA_A$ receptors consist of $2\alpha_1$-, $2\beta_2$-, and a single γ_2-subunit [21]. The benzodiazepine binding site occurs at the interface between the α- and γ-subunits in the same way that the GABA binding site occurs between the α- and β-subunits. Consequently, a single α_1-subunit can make up part of both the GABA and benzodiazepine binding sites [21]. The result is that the GABA and the benzodiazepine binding sites are physically linked, so that occupation of one of the sites can alter the configuration of the other site, thus allowing for the allosteric modulation of the activity of GABA at the $GABA_A$ receptor by the benzodiazepine binding site (Figure 3.1).

Benzodiazepine Agonism

The benzodiazepines are not agonists in the strict sense of the word, as their binding to the allosteric site does not directly produce an effect. Rather, binding to the benzodiazepine binding site by an agonist induces a conformational shift resulting in enhanced current in response to activation by GABA (Figure 3.2). There are

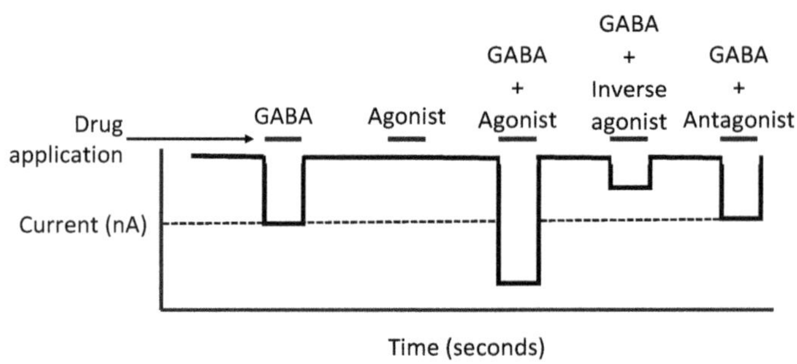

FIGURE 3.2 The effects of GABA and benzodiazepine modulators on GABA-induced current are shown schematically. When GABA interacts with its recognition site, the chloride channel opens and allows the inward flow of the negatively charged chloride ions, thus causing hyperpolarization. Benzodiazepines are positive allosteric modulators that enhance GABA activity, resulting in increased hyperpolarization evoked by GABA interacting at its recognition site. Benzodiazepines in the absence of GABA do not open the chloride channel and do not evoke any current. Compounds that act as antagonists at the benzodiazepine site neither enhance nor diminish the effect of GABA. Negative allosteric modulators, or "inverse agonists," reduce the effect of GABA on the chloride channel, resulting in a decreased level of hyperpolarization (see ref. [15]).

numerous ligands that are capable of interacting with the benzodiazepine binding site. Agonists at the benzodiazepine binding site, such as diazepam, potentiate currents induced by GABA and are positive modulators. The “inverse agonists” act in a manner opposite of benzodiazepine agonists, inhibiting GABA-induced currents, and are considered to be negative modulators of the $GABA_A$ receptor [22, 23]. The antagonists simply bind to the benzodiazepine site, without producing an effect on currents, and are thus “zero” modulators [24].

Early studies performed with isolated cultured embryonic chick spinal cord neurons measured conduction changes indicative of an increase in permeability to chloride ions in response to the iontophoretic or injected presence of GABA [25]. Chlordiazepoxide produced a concentration-dependent augmentation of the responses to GABA and enhanced synaptic potentials mediated by GABA [25]. Importantly, chlordiazepoxide alone did not have any direct effect on the neurons. Studies that show that diazepam binds to selective sites in the brain, but that it is not displaced by GABA or by GABA antagonists, suggested that benzodiazepines bind to a modulating site that can modify the coupling between occupation of the receptor by GABA and the opening of the chloride channels [25]. Electrophysiologic studies using tissue-cultured mammalian spinal cord neurons showed that chlordiazepoxide and diazepam produced a dose-dependent augmentation of the responses elicited by GABA [26]. Moreover, these studies also showed that benzodiazepines act by modulating the effect of GABA at its receptor, and not by a direct action at the GABA binding site. The benzodiazepines by themselves do not open chloride channels nor do they increase conductance; rather, they increase the frequency of chloride channel opening [15]. This action contrasts with that of the barbiturates, which can elicit some opening of the chloride channels alone, and may explain the more favorable therapeutic index of benzodiazepines as compared to barbiturates.

Benzodiazepine antagonism

Because the benzodiazepines are allosteric modulators of the GABA receptor by acting through a modulatory effect that affects the activity of an endogenous ligand (i.e. GABA), there are some nuances to consider regarding the activities of agonists and antagonists. For example, when examining agonist/antagonist activities directly at a receptor site, an antagonist binding at a receptor does not activate the receptor, but blocks the activity of endogenous ligands, effectively counteracting the effect of the agonist. For example, the commonly used β-blockers block the effect of circulating epinephrine, thus preventing increased blood pressure and heart rate. However, blocking an allosteric modulatory site could simply produce no effect at all. Thus, an antagonist at the benzodiazepine binding site may block the effect of exogenously administered benzodiazepines, but it will have no effect at all on the ability of GABA to activate the $GABA_A$ receptor. In short, while it may block the activity of the modulator, it does not block the actual activator of the chloride channel. The compound Ro 15-1788, an imidazobenzodiazepinone derivative, is antagonist for the benzodiazepine modulatory site. Electrophysiologic studies showed that Ro 15-1788 at doses of up to 10 mg/kg i.v. had no effects on segmental dorsal root potentials, polysynaptic ventral root reflexes, and Renshaw cell responses to antidromic stimulation of the ventral roots [27]. However, Ro 15-1788 was able to prevent the responses to several benzodiazepines (i.e., meclonazepam, diazepam, midazolam) or to a non-benzodiazepine compound, zopiclone, which also acts as an agonist at the benzodiazepine binding site [27]. In contrast, the effects of phenobarbital on these spinal cord responses were not affected by Ro 15-1788, which is consistent with a separate allosteric modulatory site for the barbiturates. In addition, Ro 15-1788 blocked the reductions of spontaneous multiunit activity in several regions of the rat brain in response to midazolam, but had no effect alone or on the responses to phenobarbital [27]. Similar results were seen with regard to EEG patterns in rats. These studies showed that Ro 15-1788 blocks the effects of benzodiazepines directly at the benzodiazepine modulatory site. In a randomized clinical trial, Ro 15-1788 prevented the impairment, cognitive, psychomotor, and subjective effects of orally administered diazepam [28]. It was developed as flumazenil for the rapid reversal of conscious sedation in patients receiving benzodiazepines, and for the reversal of toxic doses of benzodiazepines [29].

A seemingly counterintuitive use of flumazenil is in the treatment of withdrawal from benzodiazepines in individuals who have become tolerant and dependent on these drugs. If we look at an analogous situation, individuals who are dependent on opiates will develop withdrawal symptoms if given the opiate antagonist naloxone. While a single bolus infusion of flumazenil (e.g., 1 mg in 5 min) produces signs of benzodiazepine withdrawal in tolerant individuals, multiple slow infusions of low doses of flumazenil actually reduce the appearance of withdrawal symptoms in individuals who use benzodiazepines chronically and then discontinue use [30]. This seemingly paradoxic use of a benzodiazepine antagonist is due to the allosteric modulatory relationship between benzodiazepines and the $GABA_A$ receptor. Chronic exposure to benzodiazepines causes alterations in the benzodiazepine modulatory site on the $GABA_A$ receptor, resulting in an uncoupling between the benzodiazepine and GABA binding sites and a loss of the allosteric enhancement of GABA by benzodiazepines, along with a possible down-regulation of benzodiazepine receptors [30, 31]. In studies with cultured neurons, this uncoupling was demonstrated by a reduced sensitivity of GABA-activated chloride channels to enhancement by benzodiazepines [31]. In cell culture studies, Sf9 cells were transfected to express the $GABA_A$ receptor, and it was found that a prolonged exposure to benzodiazepines changes the conformation of the $GABA_A$ receptor such that it favors internalization into the cell and uncoupling of the GABA and benzodiazepine binding sites [31]. This altered conformation is rapidly normalized by flumazenil, restoring the $GABA_A$ receptor to the normal condition [31]. In addition, it was also shown that chronic exposure to benzodiazepines can cause changes in the relative levels of $GABA_A$ receptor subunits, decreasing the levels of those that are sensitive to benzodiazepines with subunits that are less sensitive or insensitive to benzodiazepines (see ref. [31]). It is remarkable that these effects can all be reversed, or "normalized" by a single short exposure to the benzodiazepine antagonist flumazenil.

In a blinded, placebo-controlled clinical trial, individuals who were identified as abusing flunitrazepam ($N=18$) or lormetazepam for ≥ 9 months were treated with flumazenil or placebo infusions [32]. Patients receiving flumazenil had significantly lower withdrawal scores when compared to those on placebo. Patients in the flumazenil-treated group reported improved family relationships, attentiveness, and mood perceptions [32]. This study showed the potential for the use of a benzodiazepine antagonist in managing withdrawal from chronic use of benzodiazepines.

Reverse agonism

The testing of structural modifications made to ethyl β-carboline-3-carboxylate (β-CCE), an antagonist at the benzodiazepine recognition site, led to the discovery of inverse agonists, or negative modulators of the $GABA_A$ receptor [24]. It was found that methyl 6,7-dimethoxy-4-ethyl-β-carboline-3-carboxylate (DMCM) caused convulsions in mice and rats, as did a related compound, methyl β-carboline-3-carboxylate (β-CCM). Convulsions occurred at doses at which these compounds bound to the benzodiazepine site [24]. Convulsions were blocked by benzodiazepines and barbiturates, but also by the benzodiazepine antagonist Ro 15-1788. Moreover, while benzodiazepines enhance binding of GABA to the $GABA_A$ receptor, the negative modulators of the $GABA_A$ receptor reduced the binding affinity for GABA. These inverse agonists have preferential binding to the conformation of the $GABA_A$ receptor with a closed chloride channel, and lock the channel in a configuration that reduces the likelihood of its opening in response to GABA. Thus, the inverse agonists block the inhibitory effect of GABA on neurons [24]. Animal studies have suggested that partial inverse agonists, or negative allosteric modulators, especially those acting at selective subtypes, may be useful in managing inflammatory and neuropathic pain, cognitive impairment (e.g., Alzheimer's and Down syndromes), and recovery from stroke [15].

RECEPTOR SUBTYPE COMPOSITION MAY ALTER BENZODIAZEPINE EFFECTS

There are six isoforms of the α-subunit and three forms of the γ-subunit. The benzodiazepine binding site is formed at the interface between the α- and γ-subunits. In addition, the γ_2-subunit is required to form a classical benzodiazepine binding site; consequently, the potential combinations of α/β/γ-subunits forming a functional benzodiazepine site are somewhat limited [33]. The development of transgenic mouse models, including knockouts or point mutations of genes coding for different $GABA_A$ subunit subtypes, and the development of selective ligands for different subtypes, has allowed for some detailed examination of these different receptor subunit compositions in response to modulators [6]. The different subtypes of the α-subunits have been found to be differentially expressed in the CNS [6, 33]. An immunohistochemical survey performed in the rat brain found that the α_1-subtype is widely expressed throughout the brain, whereas the α_2-, α_3-, and α_5-subtypes had more limited distributions [33]. The α_4- and α_6-subtypes were not included, as they represent less than 10% of immunoprecipitated GABAA receptors. $GABA_A$ receptors consisting of $\alpha_1/\beta_{2,3}/\gamma_2$, $\alpha_2/\beta_{2,3}/\gamma_2$, $\alpha_3/\beta_{2,3}/\gamma_2$, and $\alpha_5/\beta_{2,3}/\gamma_2$ are the most abundant and widespread triplets found. Notably, every brain region was found to express a unique pattern of neurons expressing the different subtype combinations, underlining the complexity and remarkable diversity of GABA-mediated neurotransmission [33], which carries with it the potential for different effects of benzodiazepines at these different regions.

Animal studies performed either with transgenic mouse models or with selective ligands have shown that $GABA_A$ receptors consisting of different isoforms of the α-subunit can mediate different psychopharmacological effects of benzodiazepines. The relative abundance of the different α-subunit subtypes is illustrated in Figure 3.3. For example, mice with point mutations of the α_1-subunit were resistant to benzodiazepine-induced sedation, and benzodiazepines that have little or no affinity for recognition sites that include the α_1-subunit, such as TPA123, TPA023, and L-838,417, have an anxiolytic activity without sedation in animal models [6]. The compound MRK-409 is a partial agonist at α_2-, α_3-, and α_5-expressing receptors, but has little activity at receptors containing the α_1-subunit. This compound was anxiolytic without sedation in animal models, but showed sedation in clinical trials. It was reasoned that even a low level of activity at α_1 was sufficient to produce sedation. In contrast, TPA023B, which is a partial agonist at $GABA_A$ receptors expressing α_2 or α_3 and antagonist at α_1, was anxiolytic without sedation in clinical trials (see ref. [6]). The psychopharmacological effects of benzodiazepines at different subtypes are summarized in Table 3.2.

Addiction liability and abuse potential are believed to depend in part on the activation of reward circuitry involving mesolimbic pathways and dopaminergic mechanisms of the ventral tegmental area (VTA). Benzodiazepines disinhibit dopaminergic transmission in the VTA, which explains their addiction/abuse liability. Benzodiazepines that are partial agonists at the benzodiazepine recognition site, and

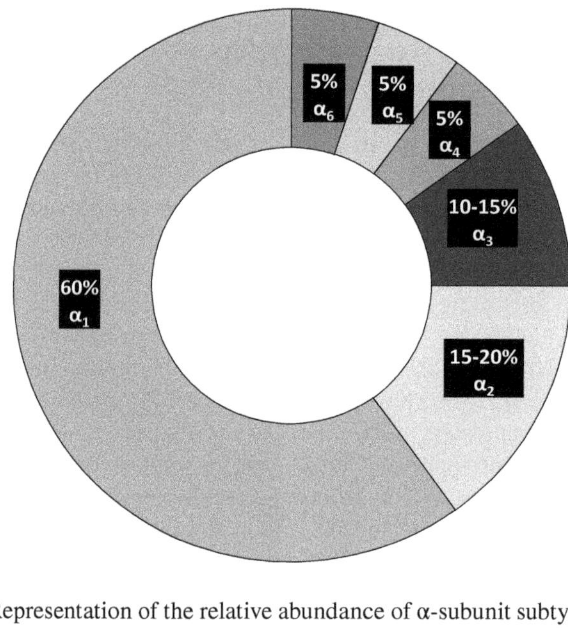

FIGURE 3.3 Representation of the relative abundance of α-subunit subtypes (see ref. [34]).

TABLE 3.2
Relative Contributions of α-Subunits to the Psychopharmacologic Properties of Benzodiazepines

Property	α1	α2	α3	α5
Sedation	Significant	None/negligible	None/negligible	None/negligible
Addiction	Significant	Minor	Minor	None/negligible
Anxiolysis	None/negligible	Significant	Significant	Minor
Myorelaxation	Minor	Significant	Minor	Minor
Anticonvulsive	Significant	Moderate	Minor	None/negligible
Amnesia	Significant	None/negligible	None/negligible	Significant

that have no activity at sites containing the α_1-subunit, such as TPA023, show very little abuse liability (see ref. [6]). Although the α_2- and α_3-subtypes are critical for the anxiolytic properties of benzodiazepines, these subtypes are also implicated with a minor role in addiction liability. It is possible, therefore, that benzodiazepines that have no or antagonistic activity at α_1, and the partial α_2 or α_3 agonists could have a limited abuse potential.

Tolerance represents a complex phenomenon, in part because it develops differently, or not at all, to different psychopharmacologic properties of benzodiazepines. For example, tolerance develops fairly rapidly to sedative and anticonvulsant effects of benzodiazepines, but not to their anxiolytic or amnesic effects [6]. Differences in the role of coupling, and of receptor down-regulation, among the receptors expressing different α-subtypes are still needed for a better understanding of this property.

SUMMARY

The benzodiazepines were met with considerable enthusiasm when they were first introduced in the 1960s and 1970s, as they were perceived to be a safer substitute for barbiturates and to be free of abuse and addiction liability. However, over time, experience showed that these optimistic projections were not really accurate, and the abuse liability of benzodiazepines became quite evident. It was also found that, like many other psychopharmaceuticals, they also could produce tolerance and dependence, with potentially dangerous withdrawal syndromes. Moreover, the safety profile that benzodiazepines enjoy is sharply reduced when they are used in combination with other sedatives. Expanding research into the pharmacology of benzodiazepines shows that their activity is due to a positive allosteric modulation of the $GABA_A$ receptor. The awareness that the $GABA_A$ receptor is composed of variations of subunit combinations, together with the discovery that the subunits themselves have different subtypes, led to investigations into the role of the subtypes in different aspects of the psychopharmacology of benzodiazepines. It seems that the possibility exists for the development of safer benzodiazepine-like drugs, with reduced, if not totally absent, abuse liability and tolerance development.

REFERENCES

1. Ban TA. The role of serendipity in drug discovery. *Dialogues Clin Neurosci.* 2006; 8(3):335–44.
2. López-Muñoz F, Alamo C, García-García P. The discovery of chlordiazepoxide and the clinical introduction of benzodiazepines: half a century of anxiolytic drugs. *J Anxiety Disord.* 2011;25(4):554–62.
3. Sternbach LH. The benzodiazepine story. *J Med Chem.* 1979;22(1):1–7.
4. Bryan J. Landmark drugs: the discovery of benzodiazepines and the adverse publicity that followed. *Pharm J.* 2009;283:305.
5. Griffin CE, Kaye AM, Bueno FR, Kaye AD. Benzodiazepine pharmacology and central nervous system-mediated effects. *Ochsner J.* 2013;13(2):214–23.
6. Cheng T, Wallace DM, Ponteri B, Tuli M. Valium without dependence? Individual GABA. *Neuropsychiatr Dis Treat.* 2018;14:1351–61.
7. Wafford KA. GABAA receptor subtypes: any clues to the mechanism of benzodiazepine dependence? *Curr Opin Pharmacol.* 2005;5(1):47–52.
8. Longo LP, Johnson B. Addiction: Part I. Benzodiazepines—side effects, abuse risk and alternatives. *Am Fam Physician.* 2000;61(7):2121–8.
9. Wick JY. The history of benzodiazepines. *Consult Pharm.* 2013;28(9):538–48.
10. Fluyau D, Revadigar N, Manobianco BE. Challenges of the pharmacological management of benzodiazepine withdrawal, dependence, and discontinuation. *Ther Adv Psychopharmacol.* 2018;8(5):147–68.
11. Schachter J, Gerstenzang, J. McFarlane tried suicide, police believe. *Los Angeles Times.* 1987.
12. Greenblatt DJ, Woo E, Allen MD, Orsulak PJ, Shader RI. Rapid recovery from massive diazepam overdose. *JAMA.* 1978;240(17):1872–4.
13. Sieghart W. GABAA receptors: ligand-gated Cl- ion channels modulated by multiple drug-binding sites. *Trends Pharmacol Sci.* 1992;13(12):446–50.
14. Sieghart W. Pharmacology of benzodiazepine receptors: an update. *J Psychiatry Neurosci.* 1994;19(1):24–9.
15. Rudolph U, Knoflach F. Beyond classical benzodiazepines: novel therapeutic potential of GABAA receptor subtypes. *Nature Reviews Drug Discovery.* 2011;10(9):685–97.
16. Mortensen M, Patel B, Smart TG. GABA potency at GABA(A) receptors found in synaptic and extrasynaptic zones. *Front Cell Neurosci.* 2011;6:1.
17. Farrant M, Nusser Z. Variations on an inhibitory theme: phasic and tonic activation of GABA(A) receptors. *Nat Rev Neurosci.* 2005;6(3):215–29.
18. Maguire JL, Stell BM, Rafizadeh M, Mody I. Ovarian cycle-linked changes in GABA(A) receptors mediating tonic inhibition alter seizure susceptibility and anxiety. *Nat Neurosci.* 2005;8(6):797–804.
19. Sigel E, Steinmann ME. Structure, function, and modulation of GABA(A) receptors. *J Biol Chem.* 2012;287(48):40224–31.
20. Ernst M, Bruckner S, Boresch S, Sieghart W. Comparative models of GABAA receptor extracellular and transmembrane domains: important insights in pharmacology and function. *Mol Pharmacol.* 2005;68(5):1291–300.
21. Sharkey LM, Czajkowski C. Individually monitoring ligand-induced changes in the structure of the GABAA receptor at benzodiazepine binding site and non-binding-site interfaces. *Mol Pharmacol.* 2008;74(1):203–12.
22. Oakley NR, Jones BJ. The proconvulsant and diazepam-reversing effects of ethyl-beta-carboline-3-carboxylate. *Eur J Pharmacol.* 1980;68(3):381–2.
23. Macdonald RL, Twyman RE, Ryan-Jastrow T, Angelotti TP. Regulation of GABAA receptor channels by anticonvulsant and convulsant drugs and by phosphorylation. *Epilepsy Res Suppl.* 1992;9:265–77.

24. Braestrup C, Schmiechen R, Neef G, Nielsen M, Petersen EN. Interaction of convulsive ligands with benzodiazepine receptors. *Science*. 1982;216(4551):1241–3.
25. Choi DW, Farb DH, Fischbach GD. Chlordiazepoxide selectively augments GABA action in spinal cord cell cultures. *Nature*. 1977;269(5626):342–4.
26. Macdonald R, Barker JL. Benzodiazepines specifically modulate GABA-mediated postsynaptic inhibition in cultured mammalian neurones. *Nature*. 1978;271(5645):563–4.
27. Polc P, Laurent JP, Scherschlicht R, Haefely W. Electrophysiological studies on the specific benzodiazepine antagonist Ro 15-1788. *Naunyn Schmiedebergs Arch Pharmacol*. 1981;316(4):317–25.
28. Darragh A, Lambe R, Kenny M, Brick I, Taaffe W, O'Boyle C. RO 15-1788 antagonises the central effects of diazepam in man without altering diazepam bioavailability. *Br J Clin Pharmacol*. 1982;14(5):677–82.
29. Reversal of central benzodiazepine effects by flumazenil after intravenous conscious sedation with diazepam and opioids: report of a double-blind multicenter study. The flumazenil in intravenous conscious sedation with diazepam multicenter study group II. *Clin Ther*. 1992;14(6):910–23.
30. Quaglio G, Pattaro C, Gerra G, Mathewson S, Verbanck P, Des Jarlais DC, et al. High dose benzodiazepine dependence: description of 29 patients treated with flumazenil infusion and stabilised with clonazepam. *Psychiatry Res*. 2012;198(3):457–62.
31. Ali NJ, Olsen RW. Chronic benzodiazepine treatment of cells expressing recombinant GABA(A) receptors uncouples allosteric binding: studies on possible mechanisms. *J Neurochem*. 2001;79(5):1100–8.
32. Gerra G, Marcato A, Caccavari R, Fertonani-Affini G, Fontanesi B, Zaimovic A, et al. Effectiveness of flumazenil (RO 15-1788) in the treatment of benzodiazepine withdrawal. *Curr Ther Res Clin Exp*. 1993;54:580–7.
33. Fritschy JM, Mohler H. GABAA-receptor heterogeneity in the adult rat brain: differential regional and cellular distribution of seven major subunits. *J Comp Neurol*. 1995;359(1):154–94.
34. Lalive AL, Rudolph U, Lüscher C, Tan KR. Is there a way to curb benzodiazepine addiction? *Swiss Med Wkly*. 2011;141:w13277.

4 Benzodiazepines and Related Substances: Therapeutic Uses and Problems

Jan M. Kitzen
Kitzen Pharmaceutical Consulting

CONTENTS

INTRODUCTION

For the past 60 years, the benzodiazepines (BZDs) have been the mainstay of pharmacotherapy for numerous psychiatric and central nervous system (CNS)-related conditions ranging from anxiety disorders, sleep disorders, management of seizures, muscle spasms, and preinduction of anesthesia. Despite their important role in therapeutics, this class of drugs has also been found to be associated with many problems, including motor vehicle crashes, falls and fractures leading to emergency room visits (especially in the elderly), cognitive impairment, and the potential for drug dependence and withdrawal.[1] In addition, the BZDs are fraught with a wide range of adverse events and the potential for misuse, especially when combined with alcohol, opioids, or other CNS depressants.[1] In 2013, 13.5 million prescriptions were written for BZDs in the United States alone.[2] During the period 2014–2016, BZDs were prescribed at approximately 66 million office-based physician visits.[3] Visit rates were higher for women than men and increased with age (see Olfson et al. for a detailed breakdown of BZD usage for specific age groups).[4] BZD use ranged from 5.7% to 6.8% for men aged 60–80 and from 10.1% to 12% for women in the same age range. It is also noteworthy that at one-third (35%) of the visits where BZDs were prescribed, an opioid was also co-prescribed.[3]

Another class of drugs similar to BZDs, but differing in chemical structure, are known as the Z-drugs since their generic names begin with the letter "z" (zolpidem, zaleplon, zopiclone, and eszopiclone, the S enantiomer of R,S (±)-zopiclone). These drugs were developed specifically for the management of insomnia, and their half-lives are much shorter than most of the BZDs. The prevalence of insomnia in American adults has been estimated to be about 10%–15% from the NHANES surveys between 1993 and 2007,[5] but other estimates are much higher, at 35%–50% in the American Insomnia Survey.[6] Regardless of the prevalence estimates, the number of physician visits with insomnia diagnoses increased 7-fold from 1993 to 2007 and the rate of prescription medications for insomnia increased by 30-fold in this same time period.[5]

This chapter will briefly review the types of drug–receptor actions at the BZD receptor binding sites, discuss the therapeutic uses and adverse events associated with the BZDs and Z-drugs, and also review the problems associated with concurrent use of BZDs and opioids.

TYPES OF LIGAND BINDING AT THE $GABA_A$ RECEPTOR ($GABA_AR$)

High-affinity binding sites for BZDs were first identified in rat brain tissue by Squires and Braestrup in 1977.[7] More recent studies have led to the elucidation of the GABA receptor structure and furthered our understanding of how the BZDs and Z-drugs interact with the GABA receptor to elicit their various pharmacologic effects.[8] The $GABA_ARs$ belong to the pentameric ligand-gated ion channel superfamily. GABA transmission is found at 20%–50% of all synapses and is consistently involved in the function of almost all brain circuits. $GABA_ARs$ are located inside the synaptic cleft of all GABAergic synapses, where they mediate phasic inhibition, or at extrasynaptic sites to mediate tonic inhibition. There are three types of modulatory effects on the GABA-induced current[2]:

1. *Agonists*, such as diazepam and clonazepam, potentiate the action of GABA by increasing the affinity of the $GABA_AR$ for GABA.
2. *Antagonists*, such as the prototype flumazenil, have no effect on the ionic conductance but block the effects of *agonists* and reduce the effects of $GABA_AR$-mediated function. Flumazenil is used clinically to treat BZD overdoses and reverse BZD-induced oversedation.
3. *Inverse agonists*, such as the β-carbolines, exert an inhibitory effect on GABAergic transmission, and have anxiogenic, proconvulsant, and myoclonic pharmacological actions.

Although there can be as many as 19 possible combinations of $GABA_AR$ subtypes, the majority of $GABA_A$ receptors are comprised of two α-, two β-, and one γ-subunits.[9] Unlike GABA itself or GABA agonists, BZD-site ligands are unable to directly activate $GABA_AR$ and are only able to modulate GABA-induced currents via allosteric modulations.[9] Each receptor complex has two binding sites for two GABA molecules and one BZD molecule.[10] BZD drugs bind to the BZD receptor at the junction of the α- and γ-subunits of the pentameric complex. This causes allosteric changes in the receptor leading to an increase in the binding by GABA, which further enhances the influx of chloride ions to prolong the inhibitory current and inhibit conductance. This inhibition is translated into anxiolytic effects through the attenuation of negative emotional stimuli activating the amygdala.[11]

Strictly speaking, the action of these compounds depends upon the presence of GABA; therefore, the terms "*agonist*," "*inverse agonist*," and "*antagonist*" are more appropriately replaced with the terms "positive allosteric modulators" (PAMs), "negative allosteric modulators" (NAMs), and "null or silent allosteric modulators" (NSAMs).[2] The effects of each of these ligand types are depicted in Figure 4.1. As shown in Figure 4.1A, PAMs (*agonists*) bind to the BZD receptor binding site and cause the allosteric modulation of the receptor complex such that the binding affinity of GABA is increased and the inhibitory effects of GABA are potentiated. This increase in GABA transmission leads to more chloride ion influx resulting in hyperpolarization, and a prolongation of the inhibitory postsynaptic current (IPSC) causes a decrease in neuronal excitability resulting in several possible pharmacologic effects

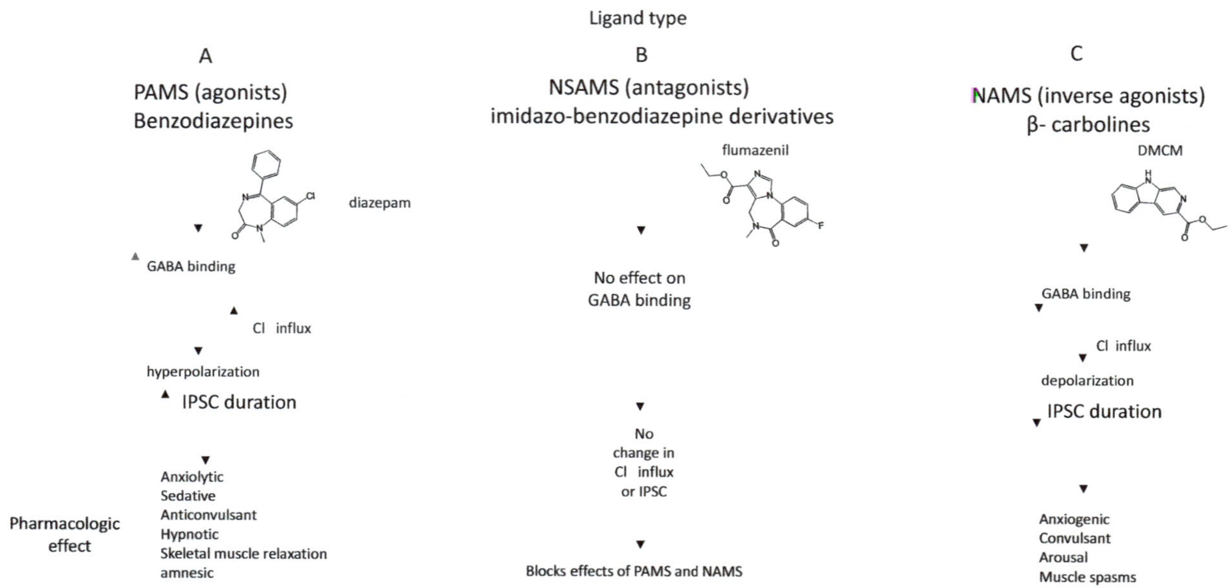

Abbreviations: GABA = Gamma-aminobutyric acid; IPSC = inhibitory postsynaptic current; PAMS: positive allosteric modulators; NSAMS: null or silent allosteric modulators; NAMS: negative allosteric modulators; DMCM: Methyl 6,7-dimethoxy-4-ethyl-beta-carboline-3-carboxylate Cl = chloride ion

FIGURE 4.1 Depiction of the three different types of ligand–$GABA_A$ receptor binding that take place with BZD and BZD-like drugs. A. Agonist interaction typified by the diazepam molecule, which acts as an agonist or a PAMs at the $GABA_A$ receptor complex. This type of binding causes an increase in the affinity of the ion-gated channel for GABA, which then leads to a greater influx of chloride ions resulting in hyperpolarization and a prolongation of the inhibitory postsynaptic current. Depending upon the specific subunit composition of the pentameric GABA receptor complex, different pharmacologic effects may occur, as shown in the lower-left portion of the figure. **B**. When a drug is able to bind to the $GABA_A$ receptor complex, but has no intrinsic activity in affecting the receptor complex (no effect on chloride influx), it is referred to as a null or silent allosteric modulator and its main effect is to block the binding of an agonist, such as any BZD or inverse agonist as shown in **C**. The net effect is an inhibition of the agonist or inverse agonist pharmacologic effects. This type of drug is also referred to as an antagonist. **C**. A third type of ligand–receptor interaction occurs when a molecule is able to bind to the $GABA_A$ receptor complex, and through a negative allosteric modulation, it inhibits the binding of GABA to the receptor complex. This leads to a decrease in chloride influx, resulting in a decrease in the inhibitory postsynaptic current. The postsynaptic neuron is in a state of partial depolarization causing it to be more excitable. This causes excitatory types of effects as shown in the lower portion of the figure.

such as anxiolytic, sedative, anticonvulsant activity, hypnotic effects, skeletal muscle relaxation, and anterograde amnesia. The NSAMs (*antagonists*), such as flumazenil (Figure 4.1B), exert no effect on chloride conductance, but they block the binding of *agonists* (PAMs) and *inverse agonists* (NAMs). The prototype of this group, flumazenil, has received marketing approval by the United States Food and Drug Administration (FDA) for reversing the various pharmacologic effects of the BZD agonists. As shown in Figure 4.1C, the NAMs-type ligands produce the opposite effects of PAMs and allosterically inhibit GABA binding, leading to a decrease in chloride influx, causing a state of partial depolarization resulting in increased excitability (decreased IPSC). The pharmacologic effects of these *inverse agonists* are characterized by anxiogenic, proconvulsant, and myotonic effects.[2]

INTERACTIONS OF ENDOGENOUS SUBSTANCES (ENDOZEPINES, EDZS) WITH THE $GABA_A$ RECEPTOR

In the mid-1970s, the discovery of the enkephalins and endorphins as endogenous substances capable of binding to opiate receptors[12,13] generated an interest to search for possible endogenous ligands of the BZD receptor. By 1983, Guidotti et al. had discovered an endogenous protein, isolated from rat brain tissue, and demonstrated that it was capable of inhibiting the binding of ^{3}[H]-diazepam. After purification to homogeneity, the protein was found to consist of 105 amino acid residues, had a molecular weight of approximately 11,000 D, and caused anxiety-like behavior when administered to rats via intracerebroventricular (i.c.v.) injection. This protein was named "diazepam binding inhibitor" (DBI), and the anxiogenic effects of DBI were blocked by the BZD receptor antagonist flumazenil. DBI was also found to be highly concentrated in brain tissue (10–25 μM) and was only found in trace amounts in peripheral organs such as liver, kidney, and spleen.[8] DBI is also known as acyl-CoA binding protein (ACBP).[14]

DBI can be cleaved by proteolytic enzymes to yield several biologically active products, including triakontatetraneuropeptide (DBI_{17-50}, TTN) and the octadecaneuropeptide DBI_{33-50} (ODN). The DBI gene is primarily expressed in astrocytes in brain tissue, and these peptide fragments are abundant in the olfactory bulb, hypothalamus, and hippocampus.[2] TTN and ODN exert a range of biological effects on anxiety-related behaviors as well as food consumption, neuroprotection, and hormonal secretions.[2] The majority of evidence suggests that the action of DBI peptides at the $GABA_AR$ is mediated by NAM activity and the exogenous application of DBI to cultured neurons reduces synaptic inhibition.[14] ODN is known to increase aggressive behavior in mice, while central administration of ODN in both rats and mice induces anxiety-like responses in several behavioral tests, which has resulted in the term "anxiety peptide" as another way to designate this peptide.[15]

BZD-like molecules have been isolated from the brains of rat, mouse, bovine brain, and humans never exposed to BZD molecules. These compounds may have originated from food as they have been isolated from milk, numerous plants, and various microorganisms.[2] DBI is widely distributed in human brain with the highest concentrations found in the arcuate nucleus of the hypothalamus, the amygdala, cerebellum, brainstem, and sacral spinal cord. This distribution pattern of DBI and peptide metabolites is consistent with their involvement in anxiety-like and stress behaviors, as well as in the control of appetite and energy homeostasis.

In addition to playing a role in anxiety and stress-related behaviors, DBI and several of its cleavage products have been implicated in the following aspects of physiological activity[2]:

- Possible role in control of body weight and energy homeostasis via neural and humoral mechanisms provides nutritional information from peripheral organs such as the gastrointestinal tract (GIT), liver, pancreas, and adipose tissue to the hypothalamus and brainstem.
- The DBI/ACBP gene is expressed in certain areas of the hypothalamus involved in the control of food intake. ODN expression occurring in the brainstem is associated with the specialized cells in the nucleus tractus solitary (NTS) and area postrema, two regions known to integrate metabolic signals originating in the GIT. I.c.v. administration of ODP in rodents decreases spontaneous food intake, while a prolonged infusion for up to 15 days decreases food consumption and body weight.
- Other mechanisms involved in modulating body weight and energy balance include such actions as modulation of DBI gene expression by fasting, hormones, or nutrients, and by targeting of anorexigenic and orexigenic circuits in hypothalamic leptin-sensitive neurons.

Distribution in Other Tissues

Despite the observations of Guidotti et al., who noted that DBI was only found in trace amounts in peripheral organs,[8] more recent studies have identified EDZs in a variety of peripheral tissues in several species. In humans, they have been found in cardiac tissue (right ventricle) widely distributed in the GIT (mostly stomach and small and large intestines) as well as pancreas, liver, and spleen and portions of the genitourinary tract and lung.[2,16]

Modulation of EDZ Release by Neurotransmitters, Neuropeptides, and Drugs

- Exposure of astrocytes to GABA leads to a dose-related inhibition of EDZ secretion, and this is also mimicked by the $GABA_BR$ agonist baclofen. Exposure of astrocytes to dibutyryl cAMP induces a time- and dose-dependent increase of EDZ release, consistent with this receptor's role in inhibiting the adenylyl cyclase/protein kinase pathway.[2]
- Somatostatin dose dependently reduces ODN output in cultured rat astrocytes, and the actions of somatostatin on neurons and endocrine cells are mediated by multiple G protein-coupled receptor subtypes. Several psychoactive drugs have been shown to modulate EDZ production.[2]
- Long-term exposure of cultured mouse neurons to ethanol stimulates DBI expression, and chronic administration of diazepam increases DBI protein content and mRNA levels in rat cerebral cortex and cerebellum. Acute

> administration of methamphetamine inhibits DBI expression in cultured astroglial cells, while exposure of cultured neurons to morphine enhances the levels of DBI mRNAs. In addition, opiate withdrawal in mice is also associated with a significant increase in DBI mRNA in the cerebral cortex that is suppressed with naloxone administration, indicating the involvement of μ-opioid receptors in the increase in DBI mRNA levels.[2]

The broad range of observations regarding the regulation and modulation of EDZs by drugs, hormones, and other physiological substances strongly suggests that these molecules have an important role in modulating neurophysiologic functions as well as having a role in the pathophysiology of withdrawal and addiction.

PHARMACOLOGY AND THERAPEUTIC USES OF BENZODIAZEPINES

Overview

The pharmacokinetic and pharmacodynamic profiles along with their relative lipid solubility (log P) and therapeutic uses are summarized in Table 4.1. BZDs vary in terms of their lipid solubility and their elimination half-lives. Their pharmacological actions are related to the type of BZD receptor they bind to. There are two major types of BZD receptors: BZ1 and BZ2. BZ1 receptors are comprised largely of the α_1 isoform subtype and are highly concentrated in the cortex, thalamus, and cerebellum.[10] The BZ1 receptor is involved in mediating the sedative, anterograde amnesia, and anticonvulsant effects of some of the BZDs.[10] The BZ2 receptors contain the α_2-subunit isoform and are primarily located in the limbic system, motor neurons, and dorsal horn of the spinal cord. These receptors are involved in mediating the anxiolytic and skeletal muscle-relaxant actions of BZDs.[17] All BZDs are classified as Schedule IV controlled substances because of their risks of causing dependence, withdrawal, and potential for drug abuse. It should also be noted that all BZDs are considered as potentially inappropriate medications (PIMs) for use in the elderly (aged $\geq$ 65 years) according to the guidelines developed by the American Geriatrics Society (AGS Beers criteria®).[18] When these drugs are necessary to use in this age group, it is advised to use the lowest effective dose for the shortest possible time.

Use in Management of Anxiety

There are several types of anxiety disorders, and BZDs have been shown to be effective in the short-term management of acute panic disorder with or without agoraphobia.[19,20] After specific phobias, panic disorder is the second most common form of anxiety disorder with a prevalence of 6% in the USA, followed by social anxiety disorder (SAD, 2.7%) and generalized anxiety disorder (GAD, 2.2%).[19] When patients with PD are started on a combined therapy of a BZD such as clonazepam plus a selective serotonin reuptake inhibitor (SSRI) such as sertraline, it is possible to achieve a rapid onset of therapeutic improvement followed by a prolonged response after tapering the BZD dose while maintaining the SSRI dose.[21] BZDs should only be used for short-term management due to their ability to cause tolerance and withdrawal.

TABLE 4.1
Pharmacokinetic Summaries for Commonly Prescribed Benzodiazepines

Drug	Oral Dose Strengths/Form	Therapeutic Uses	Protein Binding (%)	Bioavailability (%)	Lipophilicity (log *P*)[a]	T_{max} (h)	$t_{1/2}$ (h)	Major Metabolism	Comments
Alprazolam	0.25, 0.5, 1, 2 mg T[22]	Anxiety Panic disorder	80[10]	≥90[23]	2.1[24]	1–2[23]	10–15[25]	3A4[25]	Use lower doses in patients with impaired hepatic/renal function
Clonazepam	0.5, 1, 2 mg T[26] 0.125, 0.25, 0.5, 1 and 2 mg ODT	Panic disorder Seizure disorders	85[26]	90[26]	2.4[27]	1–4[26]	18–50[10]	3A (multiple isoforms)[26]	Use with caution in patients with an impaired renal function
Lorazepam	0.5, 1, 2 mg T[28]	Anxiety	85[10]	90[23]	2.4[29]	2.5[23]	10–20[23]	Non-P450 glucuronidation[25]	Hepatic or renal impairment has little effect on pharmacokinetics[10]
Oxazepam	10, 15, 30 mg C[30]	Anxiety Alcohol w/d	97[10]	≥90[23]	2.24[31]	1–5[23]	5–15[10]	Non-P450 glucuronidation[25]	
Temazepam	7.5, 15, 22.5, 30 mg C[32]	Insomnia	96[10]	≥90[23]	2.19[33]	1–2[10]	8–15[23]	Non-P450 glucuronidation[23]	May cause complex sleep behaviors similar to the Z-drugs[32]

(Continued)

TABLE 4.1 (*Continued*)
Pharmacokinetic Summaries for Commonly Prescribed Benzodiazepines

Drug	Oral Dose Strengths/Form	Therapeutic Uses	Protein Binding (%)	Bioavailability (%)	Lipophilicity (log *P*)[a]	T_{max} (h)	$t_{1/2}$ (h)	Major Metabolism	Comments
Chlordiazepoxide	5, 10, 25 mg C[34]	Anxiety; alcohol w/d	96[10]	100[35]	2.44[36]	0.5–4[10]	5–30[10]	3A4[25]	Metabolized to multiple active metabolites[10]
Diazepam	2, 5, 10 mg T[37]	Anxiety; alcohol w/d; muscle spasms	98[10]	>90[23]	2.82[38]	1–1.5[37]	25–50[37] 40–120[10b]	3A4; 2C19[37]	Plasma levels ↓30% when taken with antacids;[37] delayed onset; metabolized to multiple active products, including oxazepam, temazepam, and desmethyldiazepam[10]
Flurazepam	15, 30 mg C[39]	Insomnia	97[10]	30–60[40]	3.8[40]	0.5–1[39]	2.3[10] 47–100[10c]	P450	Metabolized to desalkylflurazepam, N-1 hydroxyethylflurazepam

C, capsules; T, tablets; ODT, orally disintegrating tablets; w/d, withdrawal.
[a] Octanol/water partition coefficient.
[b] $t_{1/2}$ for desmethyldiazepam.
[c] $t_{1/2}$ for desalkylflurazepam.

Other types of anxiety disorders include GAD and social anxiety disorder (SAD), among others, which are more effectively managed with SSRIs or SNRIs.[41] In data obtained from 2002, in the USA during that time period, BZDs were used to treat a wide range of patients (55%–94%) with anxiety disorders. In a study designed to explore the trends in BZD use for various psychological and neurological conditions, Agarwal et al. determined that the number of patient visits to primary care providers for anxiety and depression increased from 26.6% in 2003 to 33.5% by 2015 (OR: 1.43; 95% CI = 1.05–1.95; $p = 0.003$).[1]

Induction of Anesthesia

BZDs with ultra-short half-lives such as midazolam (half-life, $t_{1/2} = <5$) h are commonly used as premedication for induction of anesthesia.[42,43] Midazolam is ideal for this due to its anxiolytic and hypnotic properties. A 5-mg dose administered intravenously (i.v.) has an onset of action in one to two minutes, and the amnestic effects appear in about 4 min. Mental function returns to normal about four hours after administration. A particular advantage of midazolam is that at physiological pH, it is highly lipophilic ($\log P = 4.33$),[44] which promotes rapid absorption from the GIT and rapid penetration across the blood–brain barrier (BBB) after i.v. administration. Sedation occurs without loss of airway reflexes or significant cardiovascular effects. Compared to diazepam, midazolam produces less venous irritation, less postoperative drowsiness but more anterograde amnesia. Midazolam is useful for inducing sedation and as an adjunct to anesthesia for a variety of diagnostic and therapeutic procedures. Midazolam does cause some respiratory depression and can significantly lower diastolic and systolic BP, causing baroreceptor reflex-induced increases in the HR. These hemodynamic and respiratory changes are within the normal realm of other anesthetic drugs such as thiopental and are reversible upon dissipation of drug action.[43]

Skeletal Muscle Relaxation

The skeletal muscle-relaxant and anxiolytic effects of BZDs are mediated via α_2-containing $GABA_A$ receptors in the spinal cord and motor neurons, and also partially through α_3-containing $GABA_A$ receptor complexes.[17,45] Not all of the BZDs are indicated for skeletal muscle relaxation, possibly related to lower affinities for the α_2-subunit, causing a lack of specificity. The α_2-subunit receptor complexes are highly expressed in these areas and the cerebral cortex and hippocampus, and the receptor complexes in the motor neurons are the ones primarily involved in muscle relaxation, while those in the cortex and hippocampus are involved in mediating anxiolytic activity.[45] The muscle-relaxant effects of diazepam require higher doses than for anxiolytic effects probably due to the requirement for greater receptor occupancy to achieve this pharmacological effect.[45] The recommended oral dose for diazepam for management of skeletal muscle spasms is 2–10 mg, three to four times daily.[10] As shown in Table 4.1, diazepam is metabolized to three active metabolites (oxazepam, temazepam, and desmethyldiazepam), which all contribute to the pharmacologic effects of diazepam. The long half-life of desmethyldiazepam (40–120 h) can lead to residual drug effects and also drug accumulation with repeated administration.[10]

Anticonvulsant Activity

BZDs are considered as the first-line treatment for acute seizures, including status epilepticus. Since chronic use is hampered by the development of tolerance, they are used only for management of epileptic seizures that do not respond to other measures.[23] Diazepam, clonazepam, and clorazepate are the three most commonly used BZDs for this purpose.[46] Clonazepam and clorazepate are administered orally, while diazepam may be administered orally, rectally, or parenterally. Midazolam is also an effective anticonvulsant, the buccal administration (10 mg) can be administered during status epilepticus, and this dose route has now made the rectal administration of diazepam obsolete.[23] Each of these BZDs has a long half-life (except midazolam) and readily penetrates the brain, as needed for efficacy in treating status epilepticus. The anticonvulsant activity of the BZDs is partially, but not entirely, mediated by the α_1-subunit of the $GABA_A$ receptor.[45]

Alcohol Withdrawal

The anxiolytic and sedative properties of BZDs make them excellent therapeutic choices for management of alcohol withdrawal syndrome (AWS), and BZDs are now considered the treatment of choice for prevention and treatment of symptoms of alcohol withdrawal.[47] Chronic alcoholism is a complex psychiatric/behavioral disorder, and a discussion of the underlying factors and pathophysiological details is beyond the scope of this chapter. Briefly, AWS is characterized by tremor, restlessness, insomnia, increased agitation, seizures, and hallucinations, among others.[47] The last and most serious phase of withdrawal symptoms occurs at about four to five days after alcohol cessation when the subject develops delirium tremens (DTs). This stage is characterized by erratic and unstable autonomic function, disorientation, and various sensory hallucinations. The mortality rate can range from 15% to 20% in the absence of appropriate medical care.[48]

Alcohol consumption over a prolonged period of time leads to changes in brain receptors and neurotransmitters, which adapt in an attempt to maintain normal homeostatic function.[47] The two most important neurotransmitters and receptor systems affected by chronic exposure to alcohol are the inhibitory transmitter GABA and the excitatory transmitter glutamate. Prolonged exposure to alcohol leads to a decrease in GABA levels as well as a decrease in the sensitivity of the GABA receptor. In addition, there is an increase in the excitatory transmitter glutamate along with an increased activity in the *N*-methyl-D-aspartate (NMDA) receptor.[48] This altered neurophysiological state leads to a decrease in inhibitory activity and an increase in excitatory activity at neuronal synapses. These changes remain in place after the abrupt discontinuation of alcohol exposure leading to the symptoms associated with alcohol withdrawal. It has been suggested that the multiple signs and symptoms of AWS may be thought of simply as the result of interruption of the continuous exposure of the CNS to alcohol itself.[47]

BZDs as a class are cross-tolerant with alcohol, and ethyl alcohol also binds to the GABA receptor to the α_1-subunit, although this site is different from the BZD binding site, which lies between the α- and γ-subunits.[11] This property of cross-tolerance is what enables the BZDs to replace ethanol in AWS protocols, although it must be done in a manner that does not lead to transfer of the addiction from alcohol to a BZD. Most of the subclasses of BZDs appear to be equally effective in managing AWS; however, the most commonly used BZD is diazepam, while lorazepam, oxazepam ($t_{1/2}$ range: 10–20 and 5–15 h, respectively), chlordiazepoxide ($t_{1/2}$ 5–30 h), and clonazepam ($t_{1/2}$ = 18–50 h) (Table 4.1) have also been found effective.[10,47,49] In order to optimize the process of withdrawal and minimize the appearance of withdrawal symptoms or transfer of dependence, several withdrawal protocols with quantifiable scales have been developed. One of the more commonly utilized assessment scales is the Clinical Institutes Withdrawal Assessment-Alcohol revised scale (CIWA-Ar).[50] This withdrawal assessment tool uses a 10-item scale and can be used to monitor withdrawal and also assess adjustment in medications accordingly in patients diagnosed to have AWS.[47] The scales range from 0 to 7 for each category, and the maximum score is 67. Scores of 0–8 denote absent or minimal withdrawal, scores of 9–15 indicate moderate withdrawal, and scores of >16 indicate severe withdrawal.

Selection of Benzodiazepine and Tapering Regimen

Although it is not the purpose here to recommend specific drugs and protocols for the treatment of AWS, since this decision must be highly individualized based on each individual patient's condition and medical status, selected guidelines based upon the findings of several meta-analyses performed primarily on inpatients in alcohol detoxification settings should be useful as a starting point. Most of these studies included placebo control groups and compared BZDs to other drugs known to be beneficial in the management of alcohol withdrawal such as anticonvulsants (carbamazepine), neuroleptics, bromocriptine, doxepin, clonidine, hydroxyzine, and propranolol, among others.

In general, the selection of a specific BZD should be guided by onset and duration of action ($t_{1/2}$), and whether or not the selected BZD is metabolized to active metabolites, which can either prolong the pharmacologic action or accumulate in patients with liver disease. Long-acting agents such as chlordiazepoxide and diazepam can provide a smooth transition away from alcohol without the risk of rebound symptoms such as seizures.[47] Chlordiazepoxide is metabolized to several active metabolites (including desmethylchlordiazepoxide ($t_{1/2}$ = 18 h), desmethyldiazepam ($t_{1/2}$ = 40–120 h), and oxazepam ($t_{1/2}$ = 5–15 h), while diazepam is also metabolized to multiple active metabolites desmethyldiazepam ($t_{1/2}$ = 40–120 h), temazepam ($t_{1/2}$ = 8–15 h), and oxazepam ($t_{1/2}$ = 5–15 h).[10]

Types of Tapering Regimens

There are three types of AWS tapering regimens used for BZDs[47]:

1. **Fixed tapering dose regimen (FTDR)**: In this regimen, fixed doses of BZDs are administered at predetermined intervals without regard to symptom severity. Initial doses are determined by the severity of the withdrawal and the time elapsed since the last alcohol intake. This is an optimal choice for outpatients who may not be able to be followed closely.
2. **Symptom-triggered regimen (STR)**: Withdrawal symptoms as assessed with the CIWA-Ar scale are used to determine the starting doses of BZDs,

and doses are adjusted according to changes in the CIWA-Ar scores that are conducted at regularly scheduled intervals. Doses depend upon the symptom severity at the time of each assessment. In general, this dosing regimen results in lower total doses of BZDs and shorter treatment duration. This is a preferred regimen since it minimizes the risk of over- or under-medicating because dosage is based upon the withdrawal symptom severity.

3. **Loading dose regimen (LDR)**: In this type of regimen, long-acting BZDs are administered to lower the risk of complications such as delirium and seizures. Because larger doses of BZDs are used and administered at short time intervals (e.g., 20 mg oral diazepam every 2 h), patients' clinical condition and severity of withdrawal symptoms need to be monitored closely before each dose.

Daeppen et al. conducted a double-blind randomized clinical trial that directly compared a fixed dose regimen ($n=61$) and symptom-triggered regimen ($n=56$) using oxazepam administered to alcohol-dependent persons in an inpatient setting for three days.[51] Study results showed that patients in the STR group required significantly lower total doses of oxazepam (95.4 ± 107.7 mg; mean, standard deviation) vs 231.4 ± 29.4 mg in the FTDR group. Treatment duration was also significantly shorter (20.0 ± 24.45 h) in the STR group compared to the duration of 62.7 ± 5.44 h in the FTDR group. The severity of withdrawal symptoms assessed by the CIWA-Ar scale tended to be higher in the STR group, decreasing from a mean of 8.1 ± 5.76 on day one to 4.2 ± 3.92 on day three vs 5.5 ± 3.69 on day one to 2.7 ± 2.71 on day three in the FTDR group. Despite these significant differences in scores between the two groups, a larger proportion of patients in the FTDR group had CIWA-Ar scores >8, indicating that a greater percentage of the patients in this group were experiencing moderate withdrawal symptoms at the beginning of the study. The overall conclusion from this study was that for many patients withdrawing from alcohol, the STR treatment plan shortens detoxification time, avoids unnecessary medications, and does not compromise safety or comfort during the withdrawal process. The use of lower dosages of BZDs in STR protocols minimizes the risks of adverse events associated with BZDs such as falls, oversedation, and respiratory depression.[48]

ADVERSE EVENTS AND OTHER PROBLEMS ASSOCIATED WITH BZDS

There are numerous side effects associated with the use of BZDs, and some of these are serious enough to cause adverse events such as falls and fractures. The most common side effects associated with BZDs are dose-related, and include drowsiness, lethargy, and fatigue at lower doses; impaired psychomotor skills, ataxia, slurred speech, diplopia, and vertigo may occur with higher doses. Also, age is an important factor since there are many physiologic changes associated with the aging process that predispose the elderly to greater risk of developing adverse events.[10,23] Many of these drugs have long half-lives or are metabolized to active metabolites with long half-lives (Table 4.1). These pharmacokinetic properties, along with their lipophilicity, can cause accumulation of drug in fatty tissue upon repeated administration, which can lead to confusion, impaired thinking, and disorientation. These symptoms of overmedication, if not properly managed, can lead to serious adverse events such as falls, fractures, head trauma, automobile accidents, and other injuries related to a lack of alertness.[52,53] These are the principal reasons that led the American Geriatrics Society to place all BZDs, regardless of half-life, on the Beers list of PIMs for use in the elderly.[18] These guidelines recommend that all BZDs be avoided in the elderly except for certain conditions such as seizure disorders, rapid eye movement sleep disorder, and severe GAD.[18]

As shown in Table 4.1, many of the BZDs are metabolized by CYP3A4 and other P450 isoforms, making them susceptible to drug interactions in the form of either decreased or increased plasma levels when taken concurrently with other drugs that may be either inducers (e.g., carbamazepine, dexamethasone, phenytoin) or inhibitors (e.g., paroxetine, venlafaxine, ketoconazole, erythromycin) of these enzymes.[54] When plasma levels are increased due to the inhibition of metabolism, the possibility of enhancing side effects such as slurred speech, sedation, ataxia, or imbalance increases.[25] BZDs that are primarily metabolized via glucuronidation (oxazepam, lorazepam, and temazepam) are less likely to have these interactions.[25]

Benzodiazepines and Z-drugs as Risk Factors for Dementia

Because of the well-known effects of BZDs on memory and cognitive function, there has been significant interest in exploring the possibility that the long-term use of BZDs may increase the risk for elderly individuals to develop dementia after several years of exposure. This is a controversial subject since numerous studies including thousands of patients, meta-analyses studies, and prospective studies have produced conflicting results, as summarized below.

Several meta-analysis studies have pooled data from case–control and prospective cohort studies with follow-up periods ranging from 4 to 25 years.[55,56] In a meta-analysis performed by Zhong et al., nearly 1,600 studies were reviewed and after selecting only those studies that met their inclusion criteria, narrowed the analysis down to six studies with sample sizes ranging from 1,063 to 25,140 for a total of 45,391 participants across all six studies with a total of 11,891 cases of dementia.[56] Results of the meta-analysis showed that ever users of BZDs had a higher risk of developing dementia (risk ratio, RR 2.03, 95% CI 1.56–2.63) compared to never users and recent users had a higher risk of dementia compared to never users (RR 1.93, 95% CI 1.33–2.79). The authors concluded that their analysis showed that the long-term use of BZDs resulted in a 1.2- to 2-fold increase in the risk of developing dementia compared to patients who never used BZDs.

Another meta-analysis that reviewed over 3,000 studies narrowed down to ten studies (four cohort and six case–control studies) consisted of 171,939 subjects and 42,025 dementia cases in patients with a mean age of >70 years.[55] Results showed that patients who had ever used a BZD had an increased risk of dementia compared to never users (rate ratio, RR 1.51, 95% CI 1.17–1.95, $p=0.002$). Long-term use of any BZD was associated with an increased risk of dementia (RR 1.21) when compared to short-term use of any BZD. The definitions of short- and long-term uses varied among the studies, but, in general, short-term use was considered to be use of any BZD for <120 days. Long-term use varied between ≥121 days and >180 days within a 1-year period to use for as long as 4 years or longer.[55] The use of long-acting BZDs, defined as any BZD with a half-life ≥ 20 h, was not strongly related to risk of dementia (RR = 1.16) when compared to patients taking short-acting (half-life < 12 h) or medium-term-acting BZDs (half-life = 12–24 h). The results of this analysis did yield evidence that BZD use is associated with the development of dementia later in life and the association was stronger in those patients using long-acting BZDs for longer durations.

It is noteworthy that each of these studies further supported the earlier findings of Gallacher et al., who showed in a prospective cohort study conducted in males (n = 1,134, 103 who had been using BZDs regularly) followed up for 22 years, showed a marked increase in the incidence of dementia (OR = 3.50, 95% CI 1.57–7.79, p = 0.002).[57]

In a retrospective case–control study that utilized data available in the Clinical Practice Research Datalink electronic records of over 17 million patients from 719 medical practices in the United Kingdom—which is representative of the population of the United Kingdom in terms of age, sex, and ethnicity—66,136 cases were identified between 2006 and 2015. After excluding cases that did not meet the inclusion criteria (aged between 65 and 99 at index date, absence of motor neuron disease, HIV, viral infections, and several other behavioral/neurological disorders), 40,770 cases were matched to over 280,000 control subjects (no dementia) on the basis of year of birth and other demographic parameters; their median age was 83 (range: 78–87); and 63% were women. Patients in each group had been exposed to either BZDs or Z-drugs for similar periods of time (7.1 years; range: 4.0–11.3 years). Logistic regression analysis was used to estimate the association between drug exposure and incidence of dementia at the index date in each group by calculating odds ratios and 95% confidence intervals before and after adjusting for confounding factors as contributing to the risk of developing dementia. Results of these multiple analyses showed that after adjusting for confounding factors, there was no significant relationship between BZD or Z-drug use and the incidence of dementia.[58]

In another comprehensive review of the literature, Ettcheto et al. examined case–control studies, cohort studies, longitudinal studies, cross-sectional studies, and meta-analyses—along with preclinical studies involving BZDs and behavioral studies coupled with $GABA_AR$ subunit binding studies—and concluded that there is still insufficient data to ensure a causal relationship between the use of BZDs and development of dementia.[59]

Despite the lack of consensus on the relationship between the long-term BZD or Z-drug use and the risk of developing dementia in elderly patients, it is clear that these drugs should be used very cautiously in the elderly and avoided if other approaches for management of insomnia are feasible such as cognitive behavioral therapy or non-BZD drugs such as ramelteon, a melatonin receptor agonist. The other major concern about the long-term use of BZDs relates to their well-known risks of tolerance, dependence, and withdrawal symptoms, which are discussed elsewhere in this book.

BZDS AND OPIOIDS

Opioids are well known for their potent analgesic properties that make them useful drugs in the management of many types of pain. The pharmacology of opioids is complex; however, one of the most important pharmacologic actions is their ability to cause respiratory depression. Morphine and all other opioids, synthetic and semisynthetic, are able to depress respiration, and evidence of depression can be detected even at low therapeutic doses, which depress all phases of respiration (rate, minute volume, and tidal exchange).[60] Opioids depress respiration mainly by causing a reduction in the responsiveness of the brainstem respiratory centers to carbon dioxide. Other mechanisms of respiratory depression are related to opioid-depressant effects on the pontine and medullary centers located in the medulla oblongata region of the brainstem. The respiratory centers regulate respiration by controlling the rate and depth of inspiration in response to subtle changes in blood oxygen and carbon dioxide concentration as well as integrating signals from peripheral chemoreceptors and stretch receptors in the lungs.[61]

BZDs and alcohol are also known to be respiratory depressants, acting via different mechanisms of action. BZDs, administered in therapeutic doses, have little effect on respiration in normal subjects; however, when administered in higher doses, they can slightly depress alveolar ventilation and cause respiratory acidosis as the result of a decrease in hypoxic drive.[46] When multiple drugs achieve the same effect (depression of respiration) via different mechanisms of action, the overall effect can be either additive or synergistic; hence the rationale for not using alcohol, BZDs, and opioids concurrently. BZDs are known to cause apnea when administered with opioids, and a person may require ventilator assistance, especially if the individual has ingested another CNS depressant such as alcohol.[46] For these reasons, the concomitant use of BZDs, alcohol and opioids place individuals at increased risk for severe respiratory depression and should be avoided.

Prescribing Patterns for BZDs and Opioids

Data from the National Institute on Drug Abuse (NIDA) show that in 2017, 11,537 overdose deaths involved BZDs and 85% of these also involved an opioid. Santo et al. conducted a review of data obtained from the National Ambulatory Medical Care Survey (NAMCS), and the data was weighted to produce national estimates of office visits.[3] A total of 87,207 patient visits during the time period 2014–2016 were reviewed. Overall, the primary diagnosis (based on the International Statistical Classification of Diseases, Tenth Revision, Clinical Modification, ICD-10-CM) in patient visits where a BZD and an opioid were co-prescribed was disease of the musculoskeletal system and connective tissue (22% [95% CI: 18–27]). Over half (54% [95% CI: 44–65]) of these were related to a spinal disorder. Diseases of the circulation such as hypertension and heart disease accounted for 10% [95% CI: 8–14] of co-prescribed drugs. Mental disorders only accounted for 8% [95% CI: 5–11], and of this 8%, anxiety disorder was diagnosed in 37% [95% CI: 25–52] of these patient visits. Mental disorder was the most common diagnosis where only a BZD was prescribed (22% [95% CI: 19–25]).

In another large review of patient visits where BZDs and opioids were co-prescribed, Ladapo et al. analyzed data on adults aged 20 years and older from the 2005–2015 NAMCS and the National Hospital Ambulatory Medical Care Survey (NHAMCS, 2005–2011), representing a total of 13,146 visits nationally.[62] These surveys capture data that denotes whether a prescription is new or ongoing. Ambulatory visits involving new prescriptions reflect the national volume of new prescriptions in ambulatory care settings. The authors estimated the number of new opioid prescriptions written for patients not receiving any other opioids, as well as whether BZDs or naloxone were included on the medication list during the same visit. During the years 2005–2010, rates of new prescriptions for opioids among adults using a BZD increased from 189 to 351 per 1,000 persons. This rate increase of 162 was statistically significant (95% CI: 29–295; $p=0.02$). After 2010, the rate decreased to 172 per 1,000 persons by 2015, and this decrease in rate of –179 (95% CI: –310 to –45; $p=0.008$) was also statistically significant. In contrast, in patients not using BZDs, the rate of new opioid prescriptions increased from 78 to 93 per 1,000 persons between the years 2005 and 2010. This rate difference of 15 (95% CI: –3 to 33; $p=0.10$) was not statistically significant, and for the time period 2010–2015, the rate continued to decrease in a statistically nonsignificant manner to 79 per 1,000 persons (rate difference: –14, 95% CI: –38 to 11; $p=0.28$). These findings show that the rate of new opioid prescriptions to patients already-prescribed BZDs is disproportionately high compared with the general population receiving opioids alone. This may account for the 4-fold increased risk of overdose for patients receiving a BZD plus an opioid compared with the prescription of opioids alone. These data led the Centers for Disease Control and Prevention (CDC) to recommend that clinicians should avoid prescribing BZDs with opioids whenever possible and seek alternate therapies for management of chronic pain. Several alternative pain management options exist, including multidisciplinary rehabilitation, acupuncture, tai chi, and yoga—all of which have been found effective in alleviating chronic low back pain.[62,63] In addition, topical nonsteroidal anti-inflammatory medications (e.g., diclofenac) are known to be effective in the management of musculoskeletal pain.[64]

Monitoring Opioid and BZD Use by Health Practitioners and the General Population

All opioids, BZDs, and Z-drugs are officially designated as controlled substances under Title 21 of the United States Code Controlled Substances Act, and their use is monitored by the Drug Enforcement Agency. With the official recognition of the opioid crisis in 2018, nationwide efforts were implemented to develop improved methods for tracking the flow of these substances from prescribing practitioners to the filling of prescriptions in pharmacies.[65] Today, many states have implemented prescription drug monitoring programs (PDMPs) to track the use of these drugs by both patients and prescribers.[66] Currently, 49 states, District of Columbia, and the United States Territory of Guam have PDMPs in place and collect data for drugs in Schedules II–IV, while some also collect data for Schedule V drugs.[65] Most of the opioids are Schedule II substances, while the BZDs and Z-drugs are Schedule IV. Many factors are considered in decisions on how to schedule a drug, but the most important criterion is the drug's liability potential for abuse. In most states, prescribing practitioners and pharmacists are required to enroll in their state's PDMP. Most states share PDMP data electronically, making it possible to track accurately the flow of controlled substances from the written prescription (or e-prescription) to the time it is filled and dispensed in the pharmacy.[65]

Psychomotor and Behavioral Effects of BZDs and Opioids

Zacny et al.[67] conducted a study in normal volunteers to explore interactive effects of alprazolam 0.5 mg with oxycodone 10 mg vs the effects of each drug alone on several aspects of psychomotor and cognitive functions as well as behavioral measures that provided information about the somatic and mood-altering effects offering insight into abuse liability. The study population consisted of 10 women and 10 men between the ages of 21 and 39 in a double-blind randomized, placebo-controlled crossover trial conducted over a period of 4 weeks in a controlled environment. After subjects were administered one of the four treatments (placebo, alprazolam 0.5 mg, oxycodone 10 mg, or alprazolam 0.5 mg plus oxycodone 10 mg), they were administered the series of tests at regular intervals for up to 5 hours post-dose.

Results showed that the combination of alprazolam and oxycodone did not produce greater liability-related subjective effects than either drug alone. Subjective ratings from two key abuse liability-related measures ("liking" and "take drug(s) again") were only significantly increased for oxycodone relative to placebo. However, in visual analog scale ratings of "dreamy," "floating," and "coasting/spaced out," the drug combination produced significant effects when compared to either drug alone. The drug combination also enhanced the psychomotor/cognitive responses of "difficulty concentrating" and "heavy or sluggish feeling" compared to either drug alone, which were also significantly different from placebo. In some of the more objective measures of drug effect, subjects' performance on the digit symbol substitution test was significantly more impaired with the drug combination relative to placebo and either drug given alone (fewer number of symbols drawn and fewer number drawn correctly). Eye–hand coordination was significantly impaired by the drug combination compared to either drug alone. Both alprazolam and oxycodone, when administered alone, induced similar degrees of exophoria (deviation of the visual axis of one eye away from that of the other), and this was significantly greater when both drugs were administered together. Another important finding in this study was that while either drug alone did not decrease arterial blood oxygen saturation rates, the drug combination slightly, but significantly, decreased this, as would be expected for drugs that can cause respiratory depression.

This study showed that at therapeutic doses of oxycodone and alprazolam, the combination of both drugs together significantly impaired several aspects of cognitive function and sensory perception such that persons would be at a greater risk of injury when performing tasks that require an optimal state of alertness (e.g., driving, operating machinery). The effects on blood oxygen saturation suggest that at higher doses or more frequent dosing of an opioid, further respiratory depression would be more likely to occur.

Opioid Overdose Deaths (OODs)

Poisoning-related deaths have become the leading cause of unintentional injury-related death in the United States with over 70,000 drug-related deaths in 2017.[68] Opioids are well known for their depressant effects on respiration. When opioids are combined with other drugs that also have respiratory depressant effects such as alcohol or BZDs, the pharmacologic actions can be additive or synergistic, and can lead to hypoxic failure and death. In 2016, over half of all opioid-related overdoses were associated with alcohol, cocaine, or BZDs.[68]

In a study conducted by Tori et al.,[68] multiple cause-of-death files from the National Vital Statistics System of the Centers for Disease Control and Prevention Wide-Ranging Online Data for Epidemiologic Research (WONDER) were utilized to study all opioid-related poisoning deaths from 1999 to 2017. All cases of drug overdoses and poisoning deaths were classified using standardized coding from the ICD-10-CM. The code for *other opioids* refers to all natural and semisynthetic opioids, including morphine, codeine, hydrocodone, and oxycodone; in this study, these opioids are referred to as prescription opioids. The code for *other synthetic narcotics* refers to fentanyl, fentanyl analogs, and tramadol, and specifically excludes methadone. These are the definitions applied to the data shown in Table 4.2. Data were

TABLE 4.2
OODs with Alcohol and Benzodiazepine Co-Involvement: Prevalence by Opioid Subtype and Sex, 2015–2017[a]

Opioid Type	No (%) of OODs		
All OODs	All (n = 122,940)	Men (n = 82,506) 67.1%	Women (n = 40,434) 32.9%
Alcohol co-involvement[b]	18,419 (15.0)	14,178 (17.2)	4,241 (10.5)
BZD co-involvement[c]	26,728 (21.7)	15,837 (19.2)	10,891 (26.9)
Prescription opioid deaths[d]			
All	41,709 (33.9)	23,932 (29.0)	17,777 (44.0)
Alcohol co-involvement[b]	5,936 (14.2)	4,123 (17.2)	1,813 (10.2)
BZD co-involvement[c]	13,905 (33.3)	7,499 (31.3)	6,406 (36.0)
Synthetic opioid deaths[e]	57,459 (46.7)	40,919 (49.6)	16,540 (40.9)
Alcohol co-involvement[b]	8,584 (14.9)	6,787 (16.6)	1,767 (10.7)
BZD co-involvement[c]	9,978 (17.4)	6,356 (15.5)	3,622 (21.9)

Source: Adapted from Tori et al.[68]

BZD, benzodiazepine; ICD-10-CM, International Statistical Classification of Diseases, Tenth Revision, Clinical Modification; OODs, opioid overdose deaths.

[a] Subtype deaths exceed all opioid deaths reported because subtypes are not mutually exclusive; all classifications are based on ICD-10-CM.

[b] Indicates the presence of either ethanol or alcohol, unspecified on death certificate.

[c] Indicates the presence of BZD on death certificate.

[d] Natural and semisynthetic opioids, including hydrocodone, oxycodone, and morphine.

[e] Excludes methadone and includes fentanyl and trazodone.

analyzed using a repeated cross-sectional analysis of OODs, and in those records where either alcohol or a BZD was also involved, data were stratified by opioid subtype as defined above.

From 1999 to 2017, a total of 399,230 poisoning deaths were reported and 204,560 occurred in persons aged 35–54 years. Approximately one-third of the deaths occurred in women and two-thirds occurred in men. During this 18-year time period, the overall mortality rates for opioids alone increased 5-fold and those involving alcohol increased 5.5-fold, while those involving BZDs increased 10.3-fold. Prescription OODs had a 15.5% BZD co-involvement in 2000, and this more than doubled by 2017 to 33.1%.

Data for the three most recent years (2015–2017) of the data collected are summarized in Table 4.2. Overall, men accounted for 67.1% of all OODs, while 32.8% occurred in women. For OODs with alcohol involvement, men had a greater percentage (17.2%) compared to women (10.5%). For OODs involving a BZD, the percentage of women (26.9%) exceeded the percentage of men (19.2%). Similar trends for alcohol and BZD co-involvement in the prescription opioid deaths and the synthetic opioid deaths were observed with men having a higher percentage of deaths when alcohol was involved (17.2% vs 10.2% and 16.6% vs 10.7%, respectively). When a BZD was co-involved in a prescription opioid or synthetic opioid death, women exceeded the men in both subtypes (36.0% vs 31.3% and 21.9% vs 15.5%, respectively).

The results of this study showed that alcohol and BZD co-involvement in OODs is common, and the prevalence of alcohol and BZDs in association with OODs has increased over the past 20 years. Although not shown in the data in Table 4.2, trend analysis revealed that the prevalence of BZDs increased from 1999 to 2010 and then declined from 2010 to 2017, while the prevalence of alcohol remained steady throughout this time period. For the final three years of the study period, BZD co-involvement accounted for 33.3% of all deaths from prescription opioids. Since alcohol and opioid access is more difficult to control than access to BZDs, which is largely under the control of physicians, one possible way to decrease BZD co-involvement is through deprescribing practices. This should be relatively attainable since the use of all prescribed opioids and BZDs can be tracked via PDMPs, as described earlier. Barriers to this approach exist, unfortunately since BZD and opioid co-prescribing rates have quadrupled from 2003 and 2015, mainly attributed to prescriptions written by primary care physicians.[1]

SUBSTANCES RELATED TO BZDS: Z-DRUGS

Three other classes of drugs are structurally different from the BZDs (see Figures 4.1A and 4.2), yet also capable of interacting with the BZD receptor binding site and acting as GABA agonists (PAMs). These drugs were developed in an effort to create new sedative–hypnotics with shorter half-lives than the BZDs with the goal of having fewer adverse events and better safety profiles than the BZDs. The chemical structures of the three Z-drugs are shown in Figure 4.2. Zopiclone is not available in the

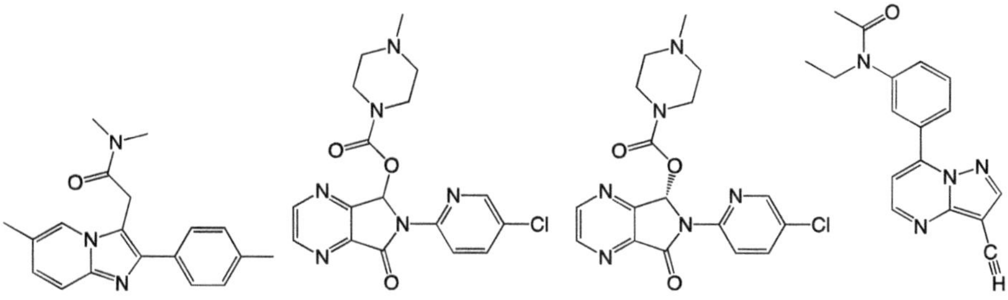

FIGURE 4.2 Chemical structures of the Z-drugs. Note the difference in chemical structures between the three different chemical classes of Z-drugs, and compare structure to diazepam in Figure 4.1A. The Z-drugs are thought to bind to the $GABA_A$ receptor complex at a site near the BZD binding site. Zopiclone is not available for use in the United States; however, the S-(+) enantiomer eszopiclone does have FDA approval for use as a hypnotic. Pharmacokinetic comparisons for each of these are summarized in Table 4.3.

USA; however, the (S)-enantiomer, eszopiclone, has been approved by the FDA for use as a sedative–hypnotic. These drugs have relatively short durations of action and are primarily used as sedative hypnotics and not as anxiolytics. Basic pharmacokinetic profiles and dosages are summarized in Table 4.3.

Mechanisms of Action for Z-drugs and Clinical Indications

The three Z-drugs available all exert their pharmacologic effects in a manner similar to the BZDs, by binding primarily to the α_1-subunit in the pentameric $GABA_A$ receptor complex at a site different from the BZD binding sites, possibly one that is allosterically coupled to the BZD receptor.[69,70] These drugs have more selectivity for the α_1-subunit compared to the BZDs, making them have a narrower spectrum of activity. Animal studies show that higher doses of Z-drugs begin to show other BZD-like effects such as anxiolytic and skeletal muscle-relaxing activity.[71] In receptor binding studies utilizing molecular biology techniques to create mutated cells with various ratios of different α-subunits in the GABA receptor pentameric complex, zolpidem was found to have the highest affinity for α_1-subunits and lower binding affinities in reassembled complexes with α_2-, α_3-, and α_5-subunits.[72,73] In a study by Hambrecht-Wiedbusch et al.,[74] α-subunit binding affinities for both zolpidem and zopiclone were compared to diazepam in preparations of brain obtained from the pontine reticular formation region of rat brain. Zolpidem, as shown previously, had the highest affinity for the α_1-subunit, and lower affinities for the α_2- and α_3-subunits. Zopiclone (and presumably eszopiclone) had an affinity profile more similar to that of diazepam (very high affinities for the α_1-, α_2-, and α_5-subunits and lower affinity for α_3-subunit), possibly explaining why eszopiclone displays greater anxiolytic activity than zolpidem when combined with SSRIs in the management of generalized anxiety disorder.[75,76]

TABLE 4.3
Pharmacokinetic Parameters and Major Metabolic Pathways for Z-drugs in Healthy Adult Subjects

Drug	Dose (mg)	Oral Bioavailability (%)	T_{max} (h)	C_{max} (ng/ml)	$t_{1/2}$ (h)	Major Metabolism Enzymes
Zolpidem IR	5[85]	65–70[70]	1.6[85]	59 (29–113)[85a]	2.6 (1.4–4.5)[85a]	3A4, 2C9, 1A2[70]
	10[85]	65–70[70]	1.6[85]	121 (58–272)[85a]	2.5 (1.4–3.8)[85a]	
Zolpidem CR	6.5[86]	65–70[70]				
	12.5[86]	65–70[70]	1.5[86]	134 (68.9–197)[86a]	2.8 (1.62–4.05)[86a]	
Zaleplon	5[70]	30[81]	0.7–1.4[70]	26.0 ± 4.4[81b]	~1[70]	Aldehyde oxidase, 3A4 (minor)[87]
	10[81]	30[81]	1.01 ± 0.17[81b]	49.0 ± 5.7[81b]	1.01 ± 0.05[81b]	
	20[81]	30[81]	1.1 ± 0.21[81b]		0.98 ± 0.03[81b]	
Eszopiclone	1–3[70]	75–80[70]	1–1.5[70]	N/A	6–7[70]	3A4, 2E1[84]
Zopiclone	7.5[79]	75[79]	1.6[79]	131 (68–197)[79a]	3.41[79]	3A4, 2C8[70]

N/A, not available.
[a] Values in parentheses are ranges.
[b] ± standard error.

Clinical Uses and Indications for Z-drugs

Insomnia

Insomnia is a sleep disorder characterized by either difficulty in falling asleep or maintaining sleep throughout the night. There are four types of symptoms associated with insomnia: difficulty in initiating sleep, difficulty in maintaining sleep, early morning awakening, and nonrestorative sleep.[6] Insomnia can be either transient (two-three nights/week), short-term (less than three weeks), or chronic (longer than four weeks).[77] In the America Insomnia Survey, Walsh et al. reported the prevalence of this condition in the USA to be between 35% and 50%.[6]

Chronic insomnia can lead to decreased productivity and lower quality of life. Other studies have shown that correlations exist between insomnia and other health problems such as obesity, diabetes, and cardiovascular disease.[78] Because of the prevalence of insomnia and the potential for adverse outcomes, it is important to have a number of treatment options available to manage this condition. In the pharmacologic management of insomnia, typical agents have consisted of BZDs, sedating antidepressants, and more recently, the non-BZD classes of drugs. The BZDs (typically, triazolam, temazepam, and flurazepam), although effective, are associated with a number of problems, including amnesia, prolonged sedation, disruption of sleep architecture as well as the risk of tolerance, physical dependence, and rebound insomnia.[70,77]

The ideal characteristics of an effective and safe sedative–hypnotic agent for the management of insomnia include a rapid onset of action, and a duration of action sufficient to maintain sleep, yet not cause residual sedation the next day.[79] The Z-drugs were developed in an effort to find new chemical entities that would be effective at inducing and maintaining sleep while having a safer adverse event profile than the BZDs. The chemical structures of the Z-drugs are shown in Figure 4.2. Each one is from a different chemical class, yet each of these drugs is able to bind to the $GABA_A$ receptor. Studies have shown that the hypnotic activity of the BZDs and the Z-drugs is due primarily to binding at the α_1-subunit of the $GABA_A$ receptor.[71]

Zolpidem

All three of the Z-drugs have FDA approvals for the short-term management of insomnia. As shown in Table 4.3, the durations of action for zolpidem and zaleplon are shorter compared to that for eszopiclone. Zolpidem is indicated for the induction and maintenance of sleep with a recommended starting dose of 10 mg in men and 5 mg in women due to gender-related differences in clearance rates. The 5 mg dose is also recommended for use in the elderly, although the recently revised Beer's criteria have added the Z-drugs to the list of drugs of PIM use in the elderly.[18] Zolpidem is available in both an immediate-release and a controlled-release form, which improves its ability to maintain sleep. As shown in Table 4.3, the 12.5-mg CR formulation of zolpidem achieves a higher C_{max} than the 10 mg IR formulation and has a slightly longer elimination phase. Other dosage forms of zolpidem have been developed, including an oral spray and sublingual tablets for a more rapid onset of action.[80] Zolpidem has been found to preserve stages 3 and 4 (nonrapid eye movement phase, or deep sleep) of sleep architecture, making it closely resemble natural sleep.[79] Zolpidem produces little next-day residual cognitive effects, but it is recommended that a second dose should not be administered during the night since next-day residual effects may occur. In general, dosage should be started with the lowest effective dose and the duration of therapy should not exceed 30 days; however, some long-term studies have administered Z-drugs for up to 12 months and found all three of the Z-drugs retained efficacy and did not lead to increases in tolerance, abuse, withdrawal effects, or new adverse events.[80] Periodic assessment of the need for long-term use is recommended.

Zaleplon

Zaleplon is an ultra-short-acting drug ($t_{1/2}$=~ 1 h, Table 4.3) that was developed to minimize the incidence of next-day residual effects on cognitive function and motor activity. It has higher affinity for the α_1-subunit and lower affinities for the α_2- and α_3-subunits in the $GABA_A$ receptor complex.[70] The lower extent of bioavailability is due to presystemic extraction.[81] The recommended dose in healthy adults is 10 mg taken immediately before retiring. A dose of 5 mg is recommended for individuals with hepatic impairment and in elderly patients aged 65 or older, although its use in the elderly population is not recommended.[18] It is primarily intended for sleep induction only, but a second dose may be taken if middle of the night awakening occurs as long as there is at least 4 h before intended awakening.[82] Unlike the other Z-drugs, the major metabolism is via aldehyde oxidase, with only minor metabolism via the cytochrome P450 enzyme system. The available data suggest that other drugs that are metabolized by CYP3A4 have only minimal effects on the pharmacokinetics of zaleplon.[79]

Eszopiclone

Eszopiclone is the S (+) stereoisomer of (R,S)-zopiclone. The S-isomer possesses essentially all of the hypnotic activity of racemic zopiclone and has a 50-fold greater affinity for the BZD receptor relative to the R (–) isomer, and shows higher affinity for the α_1- and α_3-subunits and somewhat lower affinity for the α_2- and α_5-subunits.[74,77,83] As shown in Table 4.3, the $t_{1/2}$ of eszopiclone is the longest of the three Z-drugs at approximately 6 hours and has a greater potential for residual effects at the higher doses of 2–3 mg compared to zolpidem and zaleplon. Clearance is decreased in persons ≥65 years, and lower dosage is recommended since studies found that at a 2 mg dose, this population is exposed to higher plasma concentrations for a longer period of time compared to younger adults.[77,84] Approximately 80% of an orally administered dose is bioavailable and is rapidly absorbed, and the time to peak plasma concentration is 1–1.5 h. Eszopiclone is effective in decreasing time to sleep onset and also in maintenance of sleep. The recommended starting dose in adults is 1 mg and can be increased to 2 or 3 mg if clinically necessary. Eszopiclone, 3 mg, has been studied in perimenopausal and postmenopausal women with chronic insomnia for periods as long as 6 months.[77] Study results showed that eszopiclone improved sleep efficacy measures, including a decrease in sleep latency, longer sleep maintenance, longer total sleep time, and improved quality of sleep vs placebo ($p<0.05$). Next-day functioning, physical well-being, and alertness were also improved in the eszopiclone group vs placebo.

Side Effects of Z-drugs

The most commonly observed side effects include headache, gastrointestinal upset, dizziness, amnesia, cognitive dysfunction, and ataxia. Dry mouth and unpleasant bitter taste have been reported with eszopiclone.[77] The reader should consult the product labeling for more detailed lists of side effects that are too numerous to mention here.

Adverse Events Associated with all Z-drugs

These drugs have a dangerous adverse event profile, which is why they are recommended to be avoided in the elderly according to the 2019 AGS Beers Criteria®.[18] The Z-drugs, especially zolpidem, have a peculiar adverse event of causing abnormal and complex sleep-related behaviors (parasomnias) such as sleep-driving, sleep-eating, sleep-talking, and sleep-sex, among others. These behaviors are performed in a semi-awake state of consciousness after getting out of bed during sleep and are also associated with anterograde amnesia.[82] This type of adverse event endangers the safety of the individual and possibly others, and is cause for concern. These specific adverse events led the FDA to require a boxed warning label for all Z-drugs in 2019.[88] Other serious adverse events associated with Z-drugs include residual effects post-awakening that can affect cognitive function, memory, parasomnia, and abnormal behavior. They can also cause carryover effects that can impair next-day psychomotor performances such as alertness, body balance, ability to multitask, and reaction times.[70]

BENZODIAZEPINE RECEPTOR ANTAGONIST: FLUMAZENIL

Flumazenil is an imidazobenzodiazepine derivative (Figure 4.1B) and acts as an antagonist (NSAMs) at the BZD binding site on the GABA receptor complex, thereby preventing the chloride channel-opening events and inhibiting neuronal hyperpolarization. As a result, flumazenil reverses BZD-induced effects, including sedation, psychomotor deficits, amnesia, and hypoventilation in a dose-dependent manner.[89] Flumazenil exerts suppressive effects on GABA-mediated inhibition, such as the suppression of IPSP in hippocampus and in neocortical neuronal cultures.[14] Studies have shown that flumazenil binds on the extracellular surface of the $GABA_A$ receptor at the junction between the α- and γ_2-subunits, close to the BZD binding site.[90,91] This explains the selective antagonism as flumazenil can antagonize the effects of most of the BZDs and all of the Z-drugs, but not the sedative effects of barbiturates or ethanol.[91] The drug must be administered intravenously and has been approved by the FDA for "complete or partial reversal of the sedative effects of benzodiazepines in cases where general anesthesia has been induced and/or maintained with benzodiazepines, where sedation has been produced with benzodiazepines for diagnostic and therapeutic procedures, and for the management of benzodiazepine overdose."[92] Flumazenil has a short $t_{1/2}$ of about 50 min, and the recommended dosage regimen is 3–5 mg administered at 0.5 mg/min and represents a compromise between a desirable slow awakening and the need for prompt response and a persistent effect in the overdose situation.[93] If circumstances permit, the physician may elect to use the

0.2 mg/min titration rate to slowly awaken the patient over 5–10 min. If a patient has not responded after a maximum dose of 5 mg, it is unlikely that a BZD was the cause of the overdose.[92] Sedation can recur, and cardiorespiratory monitoring is recommended.[93] Administration of flumazenil to reverse BZD-related overdose can precipitate withdrawal symptoms, including seizures and agitation, making treatment of this condition problematic, similar to the situation that may be encountered after administration of naloxone to revive an opioid overdose.

CONCLUSIONS

The BZDs and Z-drugs represent two important classes of drugs that have made significant advances in the pharmacotherapy of many neural and psychiatric conditions such as generalized anxiety disorder, panic disorder, seizure disorders, skeletal muscle relaxation, and insomnia. Despite the usefulness of these medications, their adverse event profiles and potential for addiction and dependence make them a double-edged sword. The co-prescribing of BZDs with opioids presents a potentially dangerous situation, and patients who require these drugs concurrently should be carefully monitored since the highest percentages of overdose-related deaths occur when these two types of drugs are taken together. Also, the use of BZDs and Z-drugs are not recommended for use in persons over the age of 65 since these individuals are at a higher risk of potentially serious adverse events such as falls and fractures. When these drugs are used properly, they can provide patients valuable therapeutic benefits in the management of their medical conditions for which they were prescribed. Long-term use of any of these drugs is not recommended, and, when possible, nonpharmacologic adjunctive therapies should be utilized for optimal long-term management of chronic conditions.

REFERENCES

1. Agarwal SD, Landon BE. Patterns in outpatient benzodiazepine prescribing in the United States. *JAMA Network Open*. (2019); 2(1): e187399.
2. Tonon MC, Vaudry H, Chuquet J, et al. Endozepines and their receptors: Structure, functions and pathophysiological significance. *Pharmacol Ther.* (2020); 208: 107386.
3. Santo LRP, Ashman JJ. (2020) Physician office visits at which benzodiazepines were prescribed: Findings from 2014–2016 National ambulatory medical care survey US Dept of Health and Human Services, National Center for Health Statistics 1–16 https://www.cdc.gov/nchs/data/nhsr/nhsr137-508.pdf
4. Olfson M, King M, Schoenbaum M. Benzodiazepine use in the United States. *JAMA Psychiatry*. (2015); 72(2): 136–142.
5. Bertisch SM, Herzig SJ, Winkelman JW, et al. National use of prescription medications for insomnia: NHANES 1999–2010. *Sleep*. (2014); 37(2): 343–349.
6. Walsh JK, Coulouvrat C, Hajak G, et al. Nighttime insomnia symptoms and perceived health in the America Insomnia Survey (AIS). *Sleep*. (2011); 34(8): 997–1011.
7. Squires RF, Brastrup C. Benzodiazepine receptors in rat brain. *Nature*. (1977); 266(5604): 732–734.
8. Guidotti A, Forchetti CM, Corda MG, et al. Isolation, characterization, and purification to homogeneity of an endogenous polypeptide with agonistic action on benzodiazepine receptors. *Proc Natl Acad Sci U S A*. (1983); 80(11): 3531–3535.

9. Sieghart W, Ramerstorfer J, Sarto-Jackson I, et al. A novel GABA(A) receptor pharmacology: drugs interacting with the alpha(+) beta(-) interface. *Br J Pharmacol.* (2012); 166(2): 476–485.
10. Griffin CE, 3rd, Kaye AM, Bueno FR, et al. Benzodiazepine pharmacology and central nervous system-mediated effects. *Ochsner J.* (2013); 13(2): 214–223.
11. Chen X, van Gerven J, Cohen A, et al. Human pharmacology of positive GABA-A subtype-selective receptor modulators for the treatment of anxiety. *Acta Pharmacol Sin.* (2019); 40(5): 571–582.
12. Hughes J, Smith TW, Kosterlitz HW, et al. Identification of two related pentapeptides from the brain with potent opiate agonist activity. *Nature.* (1975); 258(5536): 577–580.
13. Simantov R, Snyder SH. Morphine-like peptides in mammalian brain: isolation, structure elucidation, and interactions with the opiate receptor. *Proc Natl Acad Sci U S A.* (1976); 73(7): 2515–2519.
14. Farzampour Z, Reimer RJ, Huguenard J. Endozepines. *Adv Pharmacol.* (2015); 72147–164.
15. Bender AS, Hertz L. Octadecaneuropeptide (ODN; 'anxiety peptide') displaces diazepam more potently from astrocytic than from neuronal binding sites. *Eur J Pharmacol.* (1986); 132(2–3): 335–336.
16. Ball JA, Ghatei MA, Sekiya K, et al. Diazepam binding inhibitor-like immunoreactivity(51–70): distribution in human brain, spinal cord and peripheral tissues. *Brain Res.* (1989); 479(2): 300–305.
17. Crestani F, Löw K, Keist R, et al. Molecular targets for the myorelaxant action of diazepam. *Mol Pharmacol.* (2001); 59(3): 442–445.
18. American Geriatrics Society. 2019 updated AGS Beers Criteria® for potentially inappropriate medication use in older adults. *J Am Geriatr Soc.* (2019); 67(4): 674–694.
19. Bandelow B, Michaelis S, Wedekind D. Treatment of anxiety disorders. *Dialogues Clin Neurosci.* (2017); 19(2): 93–107.
20. Breilmann J, Girlanda F, Guaiana G, et al. Benzodiazepines versus placebo for panic disorder in adults. *Cochrane Database Syst Rev.* (2019); 3(3): Cd010677.
21. Goddard AW, Brouette T, Almai A, et al. Early coadministration of clonazepam with sertraline for panic disorder. *Arch Gen Psychiatry.* (2001); 58(7): 681–686.
22. Mylan Pharmaceuticals. (2017) Alprazolam prescribing information Drugs.com. https://www.drugs.com/pro/alprazolam.html#s-34068-34067
23. Howard P, Twycross R, Shuster J, et al. Benzodiazepines. *J Pain Symptom Manage.* (2014); 47(5): 955–964.
24. PubChem Database. (2020) Alprazolam. Accessed June 30, 2020 https://pubchem.ncbi.nlm.nih.gov/compound/2118
25. Raj A, Sheehan D. (2004) Benzodiazepines. In: *Textbook of Psychopharmacology* (Schatzberg A.F. and Nemeroff CB, eds) 3rd edn. Washington, DC: American Psychiatric Publishing, 371–389.
26. Genentech. (2016) Klonopin tablets prescribing information. South San Francisco, USA.
27. PubChem Database. (2020) Clonazepam. Accessed June 30, 2020 https://pubchem.ncbi.nlm.nih.gov/compound/2802
28. Leading Pharma. (2018) Lorazepam prescribing information Drugs.com. https://www.drugs.com/pro/lorazepam.html
29. PubChem Database. (2020) Lorazepam. Accessed June 30, 2020 https://pubchem.ncbi.nlm.nih.gov/compound/3958
30. TruPharma. (2018) Oxazepam prescribing information Drugs.com. https://www.drugs.com/pro/oxazepam.html
31. PubChem Database. (2020) Oxazepam. Accessed June 30, 2020 https://pubchem.ncbi.nlm.nih.gov/compound/4616
32. Alembic Pharmaceuticals. (2018) Temazepam prescribing information Drugs.com. Accessed June 30, https://www.drugs.com/pro/temazepam.html#s-34068-34067

33. PubChem Database. (2020) Temazepam. Accessed June 30, 2020 https://pubchem.ncbi.nlm.nih.gov/compound/5391
34. ICN Pharmaceuticals. (2005) Librium prescribing information Drugs.com. https://www.drugs.com/pro/librium.html#s-34069-34065
35. Greenblatt DJ, Divoll MK, Abernethy DR, et al. Age and gender effects on chlordiazepoxide kinetics: relation to antipyrine disposition. *Pharmacology.* (1989); 38(5): 327–334.
36. PubChem Database. (2020) Chlordiazepoxide. Accessed June 30, 2020 https://pubchem.ncbi.nlm.nih.gov/compound/2712
37. Genentech. (2017) Valium tablets prescribing information. South San Francisco.
38. PubChem Database. (2020) Diazepam. Accessed June 30, 2020 https://pubchem.ncbi.nlm.nih.gov/compound/3016
39. Mylan Pharmaceuticals. (2018) Flurazepam prescribing information Drugs.com. https://www.drugs.com/pro/flurazepam.html#s-34090-34091
40. PubChem Database. (2020) Flurazepam. Accessed June 30, 2020 https://pubchem.ncbi.nlm.nih.gov/compound/3393
41. Ravindran LN, Stein MB. The pharmacologic treatment of anxiety disorders: a review of progress. *J Clin Psychiatry.* (2010); 71(7): 839–854.
42. Greenblatt DJ, Shader RI, Divoll M, et al. Benzodiazepines: a summary of pharmacokinetic properties. *Br J Clin Pharmacol.* (1981); 11 (Suppl): 111s–116s.
43. Reves JG, Fragen RJ, Vinik HR, et al. Midazolam: pharmacology and uses. *Anesthesiology.* (1985); 62(3): 310–324.
44. Pubchem Database. (2020) Midazolam. Accessed July 4, 2020 https://pubchem.ncbi.nlm.nih.gov/compound/4192
45. Möhler H, Fritschy JM, Rudolph U. A new benzodiazepine pharmacology. *J Pharmacol Exp Ther.* (2002); 300(1): 2–8.
46. Charney D, Mihic S, Harris R. (2001) Hypnotics and Sedatives. In: *Goodman & Gilman's Pharmacological Basis for Therapeutics* (Hardman J, Limbird L, eds). 10th edn. New York: McGraw Hill, 399–427.
47. Sachdeva A, Choudhary M, Chandra M. Alcohol withdrawal syndrome: benzodiazepines and beyond. *J Clin Diagn Res.* (2015); 9(9): VE01-VE07.
48. Eberly ME, Lockwood AG, Lockwood S, et al. Outcomes after implementation of an alcohol withdrawal protocol at a single institution. *Hosp Pharm.* (2016); 51(9): 752–758.
49. Melkonian A, Patel R, Magh A, et al. Assessment of a hospital-wide CIWA-Ar protocol for management of alcohol withdrawal syndrome. *Mayo Clin Proc Innov Qual Outcomes.* (2019); 3(3): 344–349.
50. Sullivan JT, Sykora K, Schneiderman J, et al. Assessment of alcohol withdrawal: the revised clinical institute withdrawal assessment for alcohol scale (CIWA-Ar). *Br J Addict.* (1989); 84(11): 1353–1357.
51. Daeppen JB, Gache P, Landry U, et al. Symptom-triggered vs fixed-schedule doses of benzodiazepine for alcohol withdrawal: a randomized treatment trial. *Arch Intern Med.* (2002); 162(10): 1117–1121.
52. Bakken MS, Engeland A, Engesaeter LB, et al. Risk of hip fracture among older people using anxiolytic and hypnotic drugs: a nationwide prospective cohort study. *Eur J Clin Pharmacol.* (2014); 70(7): 873–880.
53. Bartlett G, Abrahamowicz M, Grad R, et al. Association between risk factors for injurious falls and new benzodiazepine prescribing in elderly persons. *BMC Fam Pract.* (2009); 101–108.
54. Cupp MJ, Tracy TS. Cytochrome P450: new nomenclature and clinical implications. *Am Fam Physician.* (1998); 57(1): 107–116.
55. He Q, Chen X, Wu T, et al. Risk of dementia in long-term benzodiazepine users: evidence from a meta-analysis of observational studies. *J Clin Neurol.* (2019); 15(1): 9–19.

56. Zhong G, Wang Y, Zhang Y, et al. Association between benzodiazepine use and dementia: a meta-analysis. *PloS one*. (2015); 10(5): e0127836.
57. Gallacher J, Elwood P, Pickering J, et al. Benzodiazepine use and risk of dementia: evidence from the caerphilly prospective study (CaPS). *J Epidemiol Community Health*. (2012); 66(10): 869–873.
58. Richardson K, Mattishent K, Loke YK, et al. History of benzodiazepine prescriptions and risk of dementia: possible bias due to prevalent users and covariate measurement timing in a nested case-control study. *Am J Epidemiol*. (2019); 188(7): 1228–1236.
59. Ettcheto M, Olloquequi J, Sánchez-López E, et al. Benzodiazepines and related drugs as a risk factor in Alzheimer's disease dementia. *Front Aging Neurosci*. (2020); 111–113.
60. Gutstein H, Akil H. (2001) Opioid Analgesics. In: *Goodman and Gilman's The Pharmacological Basis of Therapeutics* (Limbird L, Hardman J, eds). 10th edn. New York: McGraw Hill, 569–619.
61. Guyton A, Hall J. (2000) Regulation of Respiration. In: *Textbook of Medical Physiology* (Guyton A, Hall J, eds). 10th edn. Philadelphia, PA: WB Saunders Co., 474–483.
62. Ladapo JA, Larochelle MR, Chen A, et al. Physician prescribing of opioids to patients at increased risk of overdose from benzodiazepine use in the United States. *JAMA Psychiatry*. (2018); 75(6): 623–630.
63. Qaseem A, Wilt TJ, McLean RM, et al. Noninvasive treatments for acute, subacute, and chronic low back pain: a clinical practice guideline from the American college of physicians. *Ann Intern Med*. (2017); 166(7): 514–530.
64. Derry S, Conaghan P, Da Silva JA, et al. Topical NSAIDs for chronic musculoskeletal pain in adults. *Cochrane Database Syst Rev*. (2016); 4(4): Cd007400.
65. Bulloch M. (2018) The evolution of the PDMP. Accessed July 30, 2020 https://www.pharmacytimes.com/contributor/marilyn-bulloch-pharmd-bcps/2018/2007/the-evolution-of-the-pdmp
66. Anon. (2018) Prescription drug monitoring program training and technical assistance center. History of prescription drug monitoring programs. Accessed July 30, 2020 http://www.pdmpassist.org/pdf/PDMP_admin/TAG_History_PDMPs_final_20180314.pdf
67. Zacny JP, Paice JA, Coalson DW. Separate and combined psychopharmacological effects of alprazolam and oxycodone in healthy volunteers. *Drug Alcohol Depend*. (2012); 124(3): 274–282.
68. Tori ME, Larochelle MR, Naimi TS. Alcohol or benzodiazepine co-involvement with opioid overdose deaths in the United States, 1999–2017. *JAMA Network Open*. (2020); 3(4): e202361.
69. Brielmaier BD. Eszopiclone (Lunesta): a new nonbenzodiazepine hypnotic agent. *Proc (Bayl Univ Med Cent)*. (2006); 19(1): 54–59.
70. Gunja N. The clinical and forensic toxicology of Z-drugs. *J Med Toxicol*. (2013); 9(2): 155–162.
71. Sanger DJ. The pharmacology and mechanisms of action of new generation, non-benzodiazepine hypnotic agents. *CNS drugs*. (2004); 18(Suppl 1): 9–15.
72. Graham D, Faure C, Besnard F, et al. Pharmacological profile of benzodiazepine site ligands with recombinant $GABA_A$ receptor subtypes. *Eur Neuropsychopharmacol*. (1996); 6(2): 119–125.
73. Renard S, Olivier A, Granger P, et al. Structural elements of the gamma-aminobutyric acid type A receptor conferring subtype selectivity for benzodiazepine site ligands. *J Biol Chem*. (1999); 274(19): 13370–13374.
74. Hambrecht-Wiedbusch VS, Gauthier EA, Baghdoyan HA, et al. Benzodiazepine receptor agonists cause drug-specific and state-specific alterations in EEG power and acetylcholine release in rat pontine reticular formation. *Sleep*. (2010); 33(7): 909–918.
75. Krystal AD. In vivo evidence of the specificity of effects of GABA(-A) receptor modulating medications. *Sleep*. (2010); 33(7): 859–860.

76. Pollack M, Kinrys G, Krystal A, et al. Eszopiclone coadministered with escitalopram in patients with insomnia and comorbid generalized anxiety disorder. *Arch Gen Psychiatry.* (2008); 65(5): 551–562.
77. Halas CJ. Eszopiclone. *Am J Health Syst Pharm.* (2006); 63(1): 41–48.
78. Chung S, Cho SW, Jo MW, et al. The prevalence and incidence of insomnia in Korea during 2005 to 2013. *Psychiatry Investig.* (2020); 1–8.
79. Drover DR. Comparative pharmacokinetics and pharmacodynamics of short-acting hypnosedatives: zaleplon, zolpidem and zopiclone. *Clin Pharmacokinet.* (2004); 43(4): 227–238.
80. Asnis GM, Thomas M, Henderson MA. Pharmacotherapy treatment options for insomnia: a primer for clinicians. *Int J Mol Sci.* (2015); 17(1).
81. Greenblatt DJ, Harmatz JS, von Moltke LL, et al. Comparative kinetics and dynamics of zaleplon, zolpidem, and placebo. *Clin Pharmacol Ther.* (1998); 64(5): 553–561.
82. Gunja N. In the Zzz zone: the effects of Z-drugs on human performance and driving. *J Med Toxicol.* (2013); 9(2): 163–171.
83. Blaschke G, Hempel G, Müller WE. Preparative and analytical separation of the zopiclone enantiomers and determination of their affinity to the benzodiazepine receptor binding site. *Chirality.* (1993); 5(6): 419–421.
84. Sunovion Pharmaceuticals. (2019) Lunesta prescribing information Drugs.com. Accessed June 15, 2020 https://www.drugs.com/pro/lunesta.html#s-34090-34091
85. Sanofi-Aventis. (2019) Ambien prescribing information. Drugs.com. Accessed June 15, 2020 https://www.drugs.com/pro/ambien.html#s-34090-34091
86. Sanofi-Aventis. (2019) Ambien CR prescribing information Drugs.com. Accessed June 15, 2020 https://www.drugs.com/pro/ambien-cr.html#s-34090-34091
87. Pfizer Inc. (2013) Sonata prescribing information Drugs.com. Accessed June 15, 2020 https://www.drugs.com/pro/sonata.html#s-34090-34091
88. Brooks M. (2019) FDA adds boxed warning to insomnia drugs. Accessed June 30, 2020 https://www.medscape.com/viewarticle/912402?nlid=129596_913901&src=wnl_newsalrt_190430_MSCPEDIT&uac=103084MG&impID=1951351&faf=1951351
89. PubChem Database. (2020) Flumazenil. Accessed July 14, 2020 https://pubchem.ncbi.nlm.nih.gov/compound/3373
90. Kucken AM, Teissére JA, Seffinga-Clark J, et al. Structural requirements for imidazobenzodiazepine binding to GABA(-A) receptors. *Mol Pharmacol.* (2003); 63(2): 289–296.
91. Sivilotti ML. Flumazenil, naloxone and the 'coma cocktail'. *Br J Clin Pharmacol.* (2016); 81(3): 428–436.
92. Baxter Healthcare. (2018) Flumazenil injection prescribing information Drugs.com. Accessed June 22, 2020 https://www.drugs.com/pro/flumazenil-injection.html
93. An H, Godwin J. Flumazenil in benzodiazepine overdose. *CMAJ.* (2016); 188(17–18): E537.

5 Peripheral Benzodiazepine Receptors

Diana Amantea
University of Calabria

CONTENTS

INTRODUCTION

Benzodiazepines and their pharmacological congeners (e.g., Z-drugs, naturally occurring benzodiazepines, endozepines) are positive allosteric modulators of the $GABA_A$ receptor complex (Gavish and Snyder, 1980; Gielen et al., 2012), a heteropentameric ligand-gated ion channel (ionotropic)-type receptor, whose activation increases neuronal Cl^- influx (Olsen, 2018). This results in neuronal hyperpolarization, reducing cellular propensity to fire (i.e., initiate an action potential) in response to excess stimulation (analogous to anxiety or epilepsy at the clinical level). Therefore, endogenous and exogenous ligands that enhance $GABA_A$ receptor function exert beneficial effects in the context of excessive stimulation or dysfunctional inhibitory neurotransmission, such as anxiety or epilepsy (Sigel and Ernst, 2018).

Heteropentameric $GABA_A$ receptors are composed of different subunit combinations, leading to multiple receptor subtypes that display a distinct cerebral distribution and differ in their electrophysiological and pharmacological profile (Rudolph et al., 1999; Rudolph and Knoflach, 2011). The binding site for GABA is mainly formed by the β-subunit, whereas benzodiazepines predominantly interact with a "binding pocket" comprising the principal α-subunit (α1-α3 or α5) and the complementary γ-subunit (Davies et al., 1998; Kasaragod and Schindelin, 2019). Ligands at this allosteric site, acting as (full or partial) agonists, enhance GABAergic transmission, whereas antagonists, such as flumazenil, block the access of any effective ligand to this binding site (Brogden and Goa, 1991).

The first demonstration of the existence of a high-affinity binding site for benzodiazepines in the "synaptic membrane fraction" of the central nervous system was made in 1977 by radioligand binding assay, using tritiated diazepam (Möhler and Okada, 1977). This original evidence was later confirmed by a number of studies describing the localization of benzodiazepine receptors throughout the brain, both in animals and in humans (Shinotoh et al., 1986; Iyo et al., 1991; Atack, 2005). Initially, these binding sites and the related anxiolytic effects were not associated with $GABA_A$ receptors (Bræstrup and Squires, 1978), due to some mismatch in their cerebral distribution (Beaumont et al., 1978) and to the lack of anxiolytic effects of certain GABA-stimulating drugs (Cook and Sepinwall, 1975). More recent work has clarified this point, demonstrating that benzodiazepines do actually bind to an allosteric site of the $GABA_A$ receptor, resulting in potentiation of GABA inhibitory response.

An intriguing issue refers to the fact that the researchers that originally described benzodiazepine binding sites in the rat brain only focused on their cerebral function, thus marginally discussing the observation of the presence of pharmacologically different binding sites in rat kidney, liver, and lung (Möhler and Okada, 1977). Subsequent studies have highlighted the importance of these peripheral binding sites in the pharmacological effects of anxiolytic drugs and in the clinical manifestations associated with withdrawal and abuse.

DISCOVERY OF PERIPHERAL BINDING SITES FOR BENZODIAZEPINES

In 1979, the use of autoradiographic techniques allowed to demonstrate that GABA is a neurotransmitter of the vertebrate peripheral nervous system (Jessen et al., 1979; Jessen, 1981) and its presence in a number of peripheral organs was shown in the rat (Gerber and Hare, 1979; Davies and Huston, 1981; Gehlert et al., 1985). The evidence that diverse physiological actions of GABA are mediated by peripheral $GABA_A$ receptors (Krantis and Harding, 1987; Erdö et al., 1989; Tsai et al., 1993; MacNaughton et al., 1996), supported the hypothesis that also benzodiazepines may bind to receptors located outside the central nervous system and led to the identification of a novel binding site distinct from the ionotropic $GABA_A$ receptor. The peripheral benzodiazepine receptor (PBR) has been addressed using various terminologies, including "ω3 receptor," "mitochondrial benzodiazepine receptor," "tryptophan-rich sensory protein," "isoquinoline-binding protein" (IBP), or "18 kDa translocator protein" (TSPO) (Langer and Arbilla, 1988; Papadopoulos et al., 2006). In fact, this receptor has been involved in steroid biochemistry, cell proliferation, apoptosis, and modulation of immune responses. Interestingly, position emission tomography (PET) scans disclosed the widespread presence of PBR in the human body (Figure 5.1), mainly associated with mitochondria and immune cells (Endres et al., 2012). Given its abundance on the outer mitochondrial membrane of microglia activated upon brain injury, PBR expression is considered a sensitive biomarker for the detection of neuroinflammation. To this end, the isoquinoline PK11195 [1-(2-chlorophenyl)-*N*-methyl-*N*-(1-methylpropyl)-3-isoquinoline

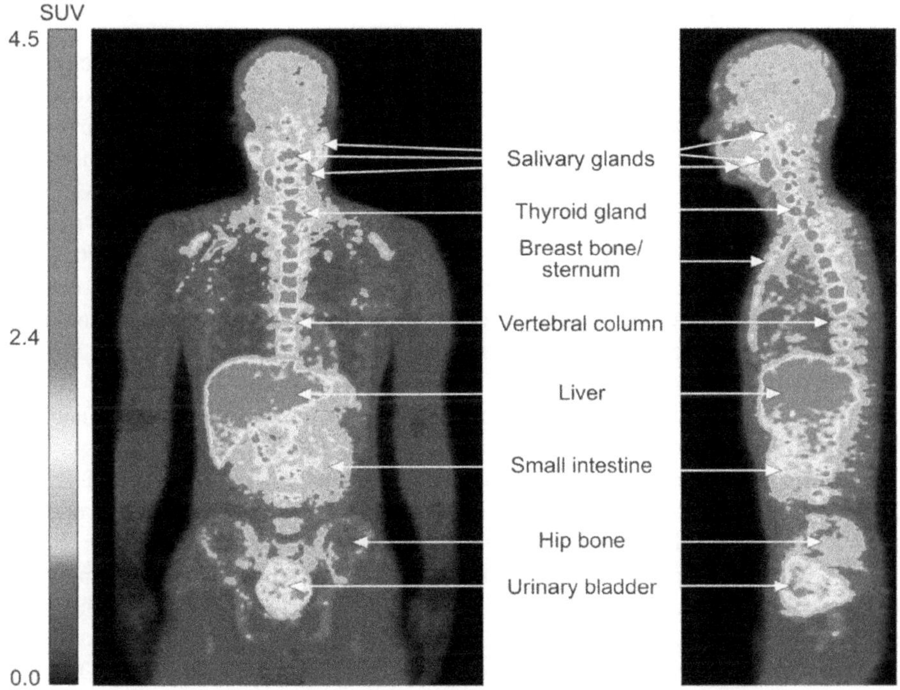

FIGURE 5.1 PET scan of a healthy human subject showing whole-body distribution of the selective PBR ligand [^{11}C]PK11195 following i.v. administration (Roivainen et al., 2011).

carboxamide] that specifically binds PBR has been tagged with [^{11}C] and is used as a tracer for in vivo receptor imaging through PET of various neuroinflammatory disorders, including Alzheimer's disease and multiple sclerosis (Guilarte, 2019; Belloli et al., 2020; Tournier et al., 2020).

The presence of PBR was detected by serendipity in peripheral tissues that were used as control during radioligand binding assays performed to characterize diazepam binding sites in the brain, the organ selected for its involvement in the anxiolytic and anticonvulsant drug effects (Braestrup and Squires, 1977). Thus, [^{3}H]diazepam binding was observed in the mitochondrial fraction of rat liver, lung, and kidney, being the latter specifically displaced by 4'-chlorodiazepam (Ro5-4864), a convulsant and anxiogenic derivative of diazepam that does not exert any effect on $GABA_A$ receptor. In fact, the PBR displays distinct pharmacological properties, since the rank order of binding affinity of most ligands (i.e., 4'-chlorodiazepam > diazepam ≥ flunitrazepam >>> clonazepam) is opposite to that reported for the central benzodiazepine receptor (CBR) (Schoemaker et al., 1983).

Apart from the selective synthetic ligands (i.e., PK11195 and Ro5-4864), various endogenous compounds interact with PBR, including cholesterol, porphyrins, peptides, such as diazepam binding inhibitor (DBI), and endozepines, namely, octadecaneuropeptide Dbi33-50 (ODN) and triakontatetraneuropeptide Dbi17-50 (TTN) derived from proteolytic processing (Costa et al., 2020).

MOLECULAR STRUCTURE

The primary coding sequence of PBR predicts a tryptophan-rich hydrophobic protein with five α-helical transmembrane domains (Korkhov et al., 2010; Jaremko et al., 2014). Actually, the supradimensional structure of the receptor includes three proteins: an 18 kDa IBP (corresponding to TSPO), a 30 kDa adenine-nucleotide transporter (ANT), and a 32 kDa voltage-dependent anion channel (VDAC) (McEnery et al., 1992). Several units of IBP associate with ANT, VDAC, and a variety of additional molecules to form a rather large complex (Woods and Williams, 1996; Costa et al., 2020). Under native conditions, murine IBP forms dimers or polymers through covalent bonds between tyrosine residues (Delavoie et al., 2003). The functional form of the receptor is believed to consist of a dimeric or polymeric state of IBP or a monomeric state interacting with other proteins (Costa et al., 2020).

Benzodiazepines bind to all three core components of the PBR complex, and several endogenous ligands of this receptor have been identified, including porphyrins, phospholipase A2, DBI, and its posttranslational product TTN (Veenman and Gavish, 2006; Berroterán-Infante et al., 2019). Thus, during benzodiazepine use, or upon its discontinuation, the physiological function of endogenous PBR ligands may be altered. In this regard, it is important to highlight that PBR ligands have distinct pharmacological properties as compared to CBR, including their insensitivity to GABA (Braestrup and Squires, 1977; Regan et al., 1980, 1981; Marangos et al., 1982; Schoemaker et al., 1983) and low binding affinity for benzodiazepines (Moingeon et al., 1984; O'Beirne and Williams, 1984; Valtier et al., 1987; Olson et al., 1988; Mihara and Fujimoto, 1989; Broaddus and Bennett, 1990; Kalk et al., 2013). Regarding the latter property, it is important to highlight that most measurements were made in conditions mimicking acute benzodiazepine exposure, whereas the effects (e.g., on receptor sensitivity or number) of chronic drug use are not contemplated.

LOCALIZATION AND FUNCTIONS

The primary coding sequence of PBR is highly conserved throughout evolution, from bacteria to humans, and the human gene (*TSPO*) is located on chromosome 22 (Riond et al., 1991; Chang et al., 1992). PBR is a highly hydrophobic 18 kDa protein that has been identified in different species and in virtually all mammalian tissues, with the highest density found in tissues containing steroid-synthesizing cells, such as adrenal glands (mainly in the cortical portion as compared to the medulla), gonads, and placenta (Fares et al., 1987; Miller, 2013; Selvaraj et al., 2015).

Moreover, PBR is localized in the salivary glands (Anholt et al., 1985), cardiovascular system (Doble et al., 1985; Veenman and Gavish, 2006), liver (Anholt et al., 1985; O'Beirne et al., 1990), and kidney (Braestrup and Squires, 1977; Schoemaker et al., 1983; Gehlert et al., 1985), and is also abundant in the brain, exceeding the density of $GABA_A$ receptors in some areas (Marangos et al., 1982; Schoemaker et al., 1983), where it is mainly located in glial cells (Dimitrova-Shumkovska et al., 2020). Initially, TSPO was considered to be specifically expressed in activated microglia and infiltrating macrophages, thus representing a useful inflammatory biomarker (Trapani et al., 2013; Venneti et al., 2013). In fact, TSPO has recently been demonstrated to be involved in redox homeostasis in microglia (Loth et al., 2020). Subsequent studies have evidenced that this receptor is also expressed in reactive astrocytes (Kuhlmann and Guilarte, 2000; Maeda et al., 2007). Under pathological conditions, the overexpression of TSPO in activated microglia and astrocytes is strictly correlated with the degree of brain injury (Chen et al., 2004; Chen and Guilarte, 2008; Dimitrova-Shumkovska et al., 2020). This peripheral receptor is mainly expressed in the outer membrane of mitochondria (Anholt et al., 1986; Antkiewicz-Michaluk et al., 1988; O'Beirne and Williams, 1988; Hirsch et al., 1989; O'Beirne et al., 1990), although it has also been found in the membrane of erythrocytes (that are devoid of mitochondria) (Olson et al., 1988), and in the endoplasmic reticulum (ER), mitochondria-associated ER membrane, and plasma membrane of microglia (Loth et al., 2020).

Activation of PBR by specific ligands results in a number of physiological responses, including inhibition of mitochondrial bioenergetics (Betlazar et al., 2020), stimulation of prolactin-stimulated mitogenesis (Laird et al., 1989), cardiovascular effects, and modulation of cell growth (Clarke and Ryan, 1980; Wang et al., 1984; Ishiguro et al., 1987; Ikezaki and Black, 1990). Moreover, PBR has also been involved in apoptosis, autophagy, oxidative stress regulation, calcium channel modulation, heme metabolism, cell differentiation and proliferation, and modulation of immune responses (Miettinen et al., 1995; Beinlich et al., 2000; Gatliff et al., 2014, 2017; Lin et al., 2014; Betlazar et al., 2020).

Overexpression of cerebral TSPO (corresponding to PBR) has been reported in various neurodegenerative conditions, such as Alzheimer's disease, Parkinson's disease, and multiple sclerosis (Rupprecht et al., 2010; Dimitrova-Shumkovska et al., 2020), and in psychiatric disorders characterized by neuroinflammation, such as schizophrenia (Doorduin et al., 2009; Bloomfield et al., 2016). Elevated receptor expression was associated with microglia and astrocytes in the inflamed brain. Pharmacological modulation of TSPO provides neuroprotection, as demonstrated in mouse models of Alzheimer's disease (Barron et al., 2013; Jung, 2020), multiple sclerosis (Daugherty et al., 2013; Ravikumar et al., 2016), and Huntington's disease (Vainshtein et al., 2015). Diverse protective mechanisms of TSPO agonists (such as PK11195, etifoxine, emapunil, and 2-Cl-MGV-1) have been postulated, including the elevated synthesis of neuroprotective steroids, anti-inflammatory and immunomodulatory effects, blockade of mitochondrial permeability transition pore opening, and reduced generation of reactive oxygen species (Dimitrova-Shumkovska et al., 2020; Jung, 2020). TSPO expression is also increased in high-grade gliomas, whereby it is one of the main PET-imaging target molecules for diagnostic purposes and has also been suggested to represent a promising therapeutic target (Bode et al., 2012; Werry et al., 2015; Roncaroli et al., 2016).

At variance with the above-described scenario, a decreased expression of PBR was observed in circulating lymphocytes, monocytes, and platelets of patients affected by anxiety-related disorders (Pini et al., 2005; Chelli et al., 2008; Abelli et al., 2010; Dell'Osso et al., 2010). Accordingly, PBR ligands have been shown to exert anxiolytic effects both in rodents and in humans (Rupprecht et al., 2009; Da Pozzo et al., 2012; Stein, 2015). Given the abundance of PBRs in various peripheral tissues and in the brain, it is intriguing to speculate that these receptors may participate in the development of undesired adverse effects associated with benzodiazepine exposure, including those related to the withdrawal syndrome.

ROLE OF PBR DURING BENZODIAZEPINE TREATMENT AND WITHDRAWAL

Prolonged use of benzodiazepines may result in tolerance to their pharmacological effects, thus requiring dose escalation and resulting in potential abuse (Licata and Rowlett, 2008). Dose escalation and dependence reflect unintentional abuse in a small percentage of patients that misuse benzodiazepines without following medical prescription (Schmitz, 2016). Although national guidelines from several countries recommend short-term (2–4 weeks) use of these drugs, they are often prescribed for long-term use, especially in the elderly, and to obtain concessionary benefits (Lader, 2011).

Upon withdrawal, benzodiazepines may cause rebound anxiety, but also a number of diverse symptoms stemming from neuronal adaptations to chronic drug exposure, including insomnia and sensory disturbances, that persist for 1–2 weeks and underlie dependence (Allison and Pratt, 2003; O'Brien, 2005). Moreover, some distressing symptoms may endure for a surprisingly long time, often unrelated to the elimination of the drug from the body, neither with the dose nor with the treatment duration. The unpleasant and dysphoric manifestations occurring upon benzodiazepine withdrawal contribute to physical dependence and sustain the need or desire to reinitiate taking the drug, not for a therapeutic need, but with the attempt to avoid this disagreeable state. Although therapeutic strategies have been suggested to be effective in tapering from benzodiazepines use, some patients report serious, protracted symptoms, complicating the efforts to successfully suspend drug treatment (Rosenberg and Chiu, 1985; Ashton, 2002). The molecular mechanisms underlying these long-lasting manifestations have not been completely clarified, and the putative direct or indirect involvement of PBR should not be underestimated.

PBR ligands stimulate neurosteroid production in the brain, independently of peripheral endocrine sources, thus promoting neurosteroid-mediated neurotransmission and providing beneficial effects in anxiety-related disorders in rodents through modulation of $GABA_A$ receptor (Da Pozzo et al., 2012). Interestingly, the non-benzodiazepine anxiolytic etifoxine has a peculiar pharmacological profile due to the fact that in addition to its modulatory effect on the $GABA_A$ receptor (Schlichter et al., 2000), it does also display a high neurosteroidogenic (e.g., *via* allopregnenolone synthesis) efficacy through the activation of PBR (Costa et al., 2017). As a consequence, etifoxine does not produce typical side effects of benzodiazepines, such as sedation, tolerance, and dependence (Stein, 2015). Intriguingly, etifoxine reduces physical signs and anxiety-like behaviors associated with ethanol withdrawal in mice, without exhibiting ataxic effects or affecting spontaneous locomotor activity (Verleye et al., 2009), strongly suggesting the involvement of PBR in the development of benzodiazepine-induced side effects. Moreover, numerous clinical studies have reported that etifoxine displays anxiolytic effects comparable with various benzodiazepines, without causing rebound anxiety or other withdrawal symptoms upon treatment interruption (Nguyen et al., 2006; Stein, 2015; Poisbeau et al., 2018; Nuss et al., 2019). Accordingly, the PBR ligand emapunil

(XBD-173 or AC-5216) exerts anxiolytic effects both in rodents and in humans without causing sedation or withdrawal symptoms, likely through newly synthetized steroids that enhance $GABA_A$ receptor function (Kita and Furukawa, 2008; Kita et al., 2009; Rupprecht et al., 2009; Nothdurfter et al., 2012).

There is a convergence of laboratory, clinical and epidemiological data reporting that benzodiazepines have different abuse liability. In particular, diazepam, lorazepam, and alprazolam display greater abuse liability than many other congeners, while oxazepam, midazolam, clorazepate, and chlordiazepoxide are relatively low in this regard (Griffiths and Wolf, 1990; Schmitz, 2016; Schippers et al., 2020). The fact that oxazepam has a favorable safety and abuse liability profile, compared with other benzodiazepine drugs, has been ascribed to its peculiar pharmacokinetic properties (e.g., slow absorption, negligible accumulation, low risk of metabolic interactions), but also to its ability to stimulate the synthesis of neurosteroids through PBR binding (Howland, 2016a, 2016b). Also, midazolam is a potent agonist of PBR, and its ability to enhance neurosteroidogenesis through the interaction with this binding site has been shown to underlie its anticonvulsant effects (Dhir and Rogawski, 2012), although there is controversial evidence on this matter (Reddy et al., 2015).

These findings highlight the crucial role of PBR in mediating diverse pharmacological and toxicological properties of benzodiazepines and their congeners, thus providing a strong background to further explore these peripheral binding sites to optimize current therapies for anxiety or epilepsy.

REFERENCES

Abelli M, Chelli B, Costa B, Lari L, Cardini A, Gesi C, Muti M, Lucacchini A, Martini C, Cassano GB, Pini S (2010) Reductions in platelet 18-kDa translocator protein density are associated with adult separation anxiety in patients with bipolar disorder. *Neuropsychobiology* 62:98–103.

Allison C, Pratt JA (2003) Neuroadaptive processes in GABAergic and glutamatergic systems in benzodiazepine dependence. *Pharmacol Ther* 98:171–195.

Anholt RRH, De Souza EB, Oster-Granite ML, Snyder SH (1985) Peripheral-type benzodiazepine receptors: Autoradiographic localization in whole-body sections of neonatal rats. *J Pharmacol Exp Ther* 233:517–526.

Anholt RRH, Pedersen PL, De Souza EB, Snyder SH (1986) The peripheral-type benzodiazepine receptor. Localization to the mitochondrial outer membrane. *J Biol Chem* 261:576–583.

Antkiewicz-Michaluk L, Guidotti A, Krueger KE (1988) Molecular characterization and mitochondrial density of a recognition site for peripheral-type benzodiazepine ligands. *Mol Pharmacol* 34:272–278.

Ashton H (2002) Benzodiazepines: How They Work; How to Withdraw. Asht Man:chapter 3 Available at: https://www.benzo.org.uk/manual/bzcha03.htm.

Atack JR (2005) The benzodiazepine binding site of GABAA receptors as a target for the development of novel anxiolytics. *Expert Opin Investig Drugs* 14:601–618.

Barron AM, Garcia-Segura LM, Caruso D, Jayaraman A, Lee JW, Melcangi RC, Pike CJ (2013) Ligand for translocator protein reverses pathology in a mouse model of Alzheimer's disease. *J Neurosci* 33:8891–8897.

Beaumont K, Chilton WS, Yamamura HI, Enna SJ (1978) Muscimol binding in rat brain: Association with synaptic GABA receptors. *Brain Res* 148:153–162.

Beinlich A, Strohmeier R, Kaufmann M, Kuhl H (2000) Relation of cell proliferation to expression of peripheral benzodiazepine receptors in human breast cancer cell lines. *Biochem Pharmacol* 60:397–402.
Belloli S, Morari M, Murtaj V, Valtorta S, Moresco RM, Gilardi MC (2020) Translation imaging in Parkinson's disease: Focus on neuroinflammation. *Front Aging Neurosci* 12:152.
Berroterán-Infante N, Tadić M, Hacker M, Wadsak W, Mitterhauser M (2019) Binding affinity of some endogenous and synthetic TSPO ligands regarding the rs6971 polymorphism. *Int J Mol Sci* 20:563.
Betlazar C, Middleton RJ, Banati R, Liu G-J (2020) The translocator protein (TSPO) in mitochondrial bioenergetics and immune processes. *Cells* 9:512.
Bloomfield PS, Selvaraj S, Veronese M, Rizzo G, Bertoldo A, Owen DR, Bloomfield MAP, Bonoldi I, Kalk N, Turkheimer F, McGuire P, De Paola V, Howes OD (2016) Microglial activity in people at ultra high risk of psychosis and in schizophrenia: An [11C]PBR28 PET brain imaging study. *Am J Psychiatry* 173:44–52.
Bode J, Veenman L, Caballero B, Lakomek M, Kugler W, Gavish M (2012) The 18 kDa translocator protein influences angiogenesis, as well as aggressiveness, adhesion, migration, and proliferation of glioblastoma cells. *Pharmacogenet Genomics* 22:538–550.
Braestrup C, Squires RF (1977) Specific benzodiazepine receptors in rat brain characterized by high-affinity [3H]diazepam binding. *Proc Natl Acad Sci U S A* 74:3805–3809.
Braestrup C, Squires RF (1978) Brain specific benzodiazepine receptors. *Br J Psychiatry* 133:249–260.
Broaddus WC, Bennett JP (1990) Peripheral-type benzodiazepine receptors in human glioblastomas: pharmacologic characterization and photoaffinity labeling of ligand recognition site. *Brain Res* 518:199–208.
Brogden RN, Goa KL (1991) Flumazenil. A reappraisal of its pharmacological properties and therapeutic efficacy as a benzodiazepine antagonist. *Drugs* 42:1061–1089.
Chang YJ, Mccabe RT, Rennert H, Budarf ML, Sayegh R, Emanuel BS, Skolnick P, Strauss JF (1992) The human "peripheral-type" benzodiazepine receptor: Regional mapping of the gene and characterization of the receptor expressed from cDNA. *DNA Cell Biol* 11:471–480.
Chelli B, Pini S, Abelli M, Cardini A, Lari L, Muti M, Gesi C, Cassano GB, Lucacchini A, Martini C (2008) Platelet 18 kDa translocator protein density is reduced in depressed patients with adult separation anxiety. *Eur Neuropsychopharmacol* 18:249–254.
Chen MK, Baidoo K, Verina T, Guilarte TR (2004) Peripheral benzodiazepine receptor imaging in CNS demyelination: Functional implications of anatomical and cellular localization. *Brain* 127:1379–1392.
Chen MK, Guilarte TR (2008) Translocator protein 18 kDa (TSPO): Molecular sensor of brain injury and repair. *Pharmacol Ther* 118:1–17.
Clarke GD, Ryan PJ (1980) Tranquillizers can block mitogenesis in 3T3 cells and induce differentiation in friend cells. *Nature* 287:160–161.
Cook L, Sepinwall J (1975) Behavior analysis of the effects and mechanisms of action of benzodiazepines. *Psychopharmacol Bull* 11:53–55.
Costa B, Cavallini C, Da Pozzo E, Taliani S, Da Settimo F, Martini C (2017) The anxiolytic etifoxine binds to TSPO Ro5-4864 binding site with long residence time showing a high neurosteroidogenic activity. *ACS Chem Neurosci* 8:1448–1454.
Costa B, Da Pozzo E, Martini C (2020) 18-kDa translocator protein association complexes in the brain: From structure to function. *Biochem Pharmacol* 177:114015.
Da Pozzo E, Costa B, Martini C (2012) Translocator protein (TSPO) and neurosteroids: Implications in psychiatric disorders. *Curr Mol Med* 12:426–442.
Daugherty DJ, Selvaraj V, Chechneva OV, Liu XB, Pleasure DE, Deng W (2013) A TSPO ligand is protective in a mouse model of multiple sclerosis. *EMBO Mol Med* 5:891–903.

Davies LP, Huston V (1981) Peripheral benzodiazepine binding sites in heart and their interaction with dipyridamole. *Eur J Pharmacol* 73:209–211.

Davies M, Bateson AN, Dunn SMJ (1998) Structural requirements for ligand interactions at the benzodiazepine recognition site of the GABA(A) receptor. *J Neurochem* 70:2188–2194.

Delavoie F, Li H, Hardwick M, Robert JC, Giatzakis C, Péranzi G, Yao ZX, Maccario J, Lacapère JJ, Papadopoulos V (2003) In vivo and in vitro peripheral-type benzodiazepine receptor polymerization: Functional significance in drug ligand and cholesterol binding. *Biochemistry* 42:4506–4519.

Dell'Osso L, Da Pozzo E, Carmassi C, Trincavelli ML, Ciapparelli A, Martini C (2010) Lifetime manic-hypomanic symptoms in post-traumatic stress disorder: Relationship with the 18 kDa mitochondrial translocator protein density. *Psychiatry Res* 177:139–143.

Dhir A, Rogawski MA (2012) Role of neurosteroids in the anticonvulsant activity of midazolam. *Br J Pharmacol* 165:2684–2691.

Dimitrova-Shumkovska J, Krstanoski L, Veenman L (2020) Diagnostic and therapeutic potential of TSPO studies regarding neurodegenerative diseases, psychiatric disorders, alcohol use disorders, traumatic brain injury, and stroke: An update. *Cells* 9:870.

Doble A, Benavides J, Ferris O, Bertrand P, Menager J, Vaucher N, Burgevin MC, Uzan A, Gueremy C, Le Fur G (1985) Dihydropyridine and peripheral type benzodiazepine binding sites: Subcellular distribution and molecular size determination. *Eur J Pharmacol* 119:153–167.

Doorduin J, De Vries EFJ, Willemsen ATM, De Groot JC, Dierckx RA, Klein HC (2009) Neuroinflammation in schizophrenia-related psychosis: A PET study. *J Nucl Med* 50:1801–1807.

Endres CJ, Coughlin JM, Gage KL, Watkins CC, Kassiou M, Pomper MG (2012) Radiation dosimetry and biodistribution of the TSPO ligand 11C-DPA-713 in humans. *J Nucl Med* 53:330–335.

Erdö SL, Mione MC, Amenta F, Wolff JR (1989) Binding of [3H]-muscimol to GABAA sites in the guinea-pig urinary bladder: Biochemical assay and autoradiography. *Br J Pharmacol* 96:313–318.

Fares F, Bar-Ami S, Brandes JM, Gavish M (1987) Gonadotropin- and estrogen-induced increase of peripheral-type benzodiazepine binding sites in the hypophyseal-genital axis of rats. *Eur J Pharmacol* 133:97–102.

Gatliff J, East D, Crosby J, Abeti R, Harvey R, Craigen W, Parker P, Campanella M (2014) TSPO interacts with VDAC1 and triggers a ROS-mediated inhibition of mitochondrial quality control. *Autophagy* 10:2279–2296.

Gatliff J, East DA, Singh A, Alvarez MS, Frison M, Matic I, Ferraina C, Sampson N, Turkheimer F, Campanella M (2017) A role for TSPO in mitochondrial Ca2+ homeostasis and redox stress signaling. *Cell Death Dis* 8:e2896–e2896.

Gavish M, Snyder SH (1980) Benzodiazepine recognition sites on GABA receptors. *Nature* 287:651–652.

Gehlert DR, Yamamura HI, Wamsley JK (1985) Autoradiographic localization of peripheral-type benzodiazepine binding sites in the rat brain, heart and kidney. *Naunyn Schmiedebergs Arch Pharmacol* 328:454–460.

Gerber JC, Hare TA (1979) Gamma-aminobutyric acid in peripheral tissue, with emphasis on the endocrine pancreas. Presence in two species and reduction by streptozotocin. *Diabetes* 28:1073–1076.

Gielen MC, Lumb MJ, Smart TG (2012) Benzodiazepines modulate GABA A receptors by regulating the preactivation step after GABA binding. *J Neurosci* 32:5707–5715.

Griffiths RR, Wolf B (1990) Relative abuse liability of different benzodiazepines in drug abusers. *J Clin Psychopharmacol* 10:237–243.

Guilarte TR (2019) TSPO in diverse CNS pathologies and psychiatric disease: A critical review and a way forward. *Pharmacol Ther* 194:44–58.

Hirsch JD, Beyer CF, Malkowitz L, Loullis CC, Blume AJ (1989) Characterization of ligand binding to mitochondrial benzodiazepine receptors. *Mol Pharmacol* 35:164–172.
Howland RH (2016a) Safety and abuse liability of oxazepam: Is this benzodiazepine drug underutilized? *J Psychosoc Nurs Ment Health Serv* 54:22–25.
Howland RH (2016b) Oxazepam for the treatment of substance abuse and depression: Is it appropriate? *J Psychosoc Nurs Ment Health Serv* 54:21–24.
Ikezaki K, Black KL (1990) Stimulation of cell growth and DNA synthesis by peripheral benzodiazepine. *Cancer Lett* 49:115–120.
Ishiguro K, Taft WC, DeLorenzo RJ, Sartorelli AC (1987) The role of benzodiazepine receptors in the induction of differentiation of HL-60 leukemia cells by benzodiazepines and purines. *J Cell Physiol* 131:226–234.
Iyo M, Itoh T, Yamasaki T, Fukuda H, Inoue O, Shinotoh H, Suzuki K, Fukui S, Tateno Y (1991) Quantitative in vivo analysis of benzodiazepine binding sites in the human brain using positron emission tomography. *Neuropharmacology* 30:207–215.
Jaremko Ł, Jaremko M, Giller K, Becker S, Zweckstetter M (2014) Structure of the mitochondrial translocator protein in complex with a diagnostic ligand. *Science* 343: 1363–1366.
Jessen KR (1981) GABA and the enteric nervous system—A neurotransmitter function? *Mol Cell Biochem* 38:69–76.
Jessen KR, Mirsky R, Dennison ME, Burnstock G (1979) GABA may be a neurotransmitter in the vertebrate peripheral nervous system [16]. *Nature* 281:71–74.
Jung ME (2020) A protective role of translocator protein in Alzheimer's disease brain. *Curr Alzheimer Res* 17:3–15.
Kalk NJ, Owen DR, Tyacke RJ, Reynolds R, Rabiner EA, Lingford-hughes AR, Parker CA (2013) Are prescribed benzodiazepines likely to affect the availability of the 18 kDa translocator protein (TSPO) in PET studies? *Synapse* 67:909–912.
Kasaragod VB, Schindelin H (2019) Structure of heteropentameric GABAA receptors and receptor-anchoring properties of gephyrin. *Front Mol Neurosci* 12:191.
Kita A, Furukawa K (2008) Involvement of neurosteroids in the anxiolytic-like effects of AC-5216 in mice. *Pharmacol Biochem Behav* 89:171–178.
Kita A, Kinoshita T, Kohayakawa H, Furukawa K, Akaike A (2009) Lack of tolerance to anxiolysis and withdrawal symptoms in mice repeatedly treated with AC-5216, a selective TSPO ligand. *Prog Neuro-Psychopharmacology Biol Psychiatry* 33:1040–1045.
Korkhov VM, Sachse C, Short JM, Tate CG (2010) Three-dimensional structure of TspO by electron cryomicroscopy of helical crystals. *Structure* 18:677–687.
Krantis A, Harding RK (1987) GABA-related actions in isolated in vitro preparations of the rat small intestine. *Eur J Pharmacol* 141:291–298.
Kuhlmann AC, Guilarte TR (2000) Cellular and subcellular localization of peripheral benzodiazepine receptors after trimethyltin neurotoxicity. *J Neurochem* 74:1694–1704.
Lader M (2011) Benzodiazepines revisited-will we ever learn? *Addiction* 106:2086–2109.
Laird HE, Gerrish KE, Duerson KC, Putnam CW, Russell DH (1989) Peripheral benzodiazepine binding sites in Nb 2 node lymphoma cells: Effects on prolactin-stimulated proliferation and ornithine decarboxylase activity. *Eur J Pharmacol* 171:25–35.
Langer SZ, Arbilla S (1988) Imidazopyridines as a tool for the characterization of benzodiazepine receptors: A proposal for a pharmacological classification as omega receptor subtypes. *Pharmacol Biochem Behav* 29:763–766.
Licata SC, Rowlett JK (2008) Abuse and dependence liability of benzodiazepine-type drugs: GABAA receptor modulation and beyond. *Pharmacol Biochem Behav* 90:74–89.
Lin R, Angelin A, Da Settimo F, Martini C, Taliani S, Zhu S, Wallace DC (2014) Genetic analysis of dTSPO, an outer mitochondrial membrane protein, reveals its functions in apoptosis, longevity, and Aβ42-induced neurodegeneration. *Aging Cell* 13:507–518.

Loth MK, Guariglia SR, Re DB, Perez J, de Paiva VN, Dziedzic JL, Chambers JW, Azzam DJ, Guilarte TR (2020) A novel interaction of translocator protein 18 kDa (TSPO) with NADPH oxidase in microglia. *Mol Neurobiol* 57:4467–4487.

MacNaughton WK, Pineau BC, Krantis A (1996) γ-Aminobutyric acid stimulates electrolyte transport in the guinea pig ileum in vitro. *Gastroenterology* 110:498–507.

Maeda J, Higuchi M, Inaji M, Ji B, Haneda E, Okauchi T, Zhang MR, Suzuki K, Suhara T (2007) Phase-dependent roles of reactive microglia and astrocytes in nervous system injury as delineated by imaging of peripheral benzodiazepine receptor. *Brain Res* 1157:100–111.

Marangos PJ, Patel J, Boulenger JP, Clark-Rosenberg R (1982) Characterization of peripheral-type benzodiazepine binding sites in brain using [3H]Ro 5-4864. *Mol Pharmacol* 22:26–32.

McEnery MW, Snowman AM, Trifiletti RR, Snyder SH (1992) Isolation of the mitochondrial benzodiazepine receptor: Association with the voltage-dependent anion channel and the adenine nucleotide carrier. *Proc Natl Acad Sci U S A* 89:3170–3174.

Miettinen H, Kononen J, Haapasalo H, Helén P, Sallinen P, Harjuntausta T, Helin H, Alho H (1995) Expression of peripheral-type benzodiazepine receptor and diazepam binding inhibitor in human astrocytomas: Relationship to cell proliferation. *Cancer Res* 55:2691–2695.

Mihara S, Fujimoto M (1989) High-affinity binding sites for PK 11195, but not for Ro5-4864, in porcine aortic smooth muscle. *Life Sci* 44:1713–1720.

Miller WL (2013) Steroid hormone synthesis in mitochondria. *Mol Cell Endocrinol* 379:62–73.

Möhler H, Okada T (1977) Benzodiazepine receptor: Demonstration in the central nervous system. *Science* 198:849–851.

Moingeon P, Dessaux JJ, Fellous R, Alberici GF, Bidart JM, Motté P, Bohuon C (1984) Benzodiazepine receptors on human blood platelets. *Life Sci* 35:2003–2009.

Nguyen N, Fakra E, Pradel V, Jouve E, Alquier C, Le Guern M-E, Micallef J, Blin O (2006) Efficacy of etifoxine compared to lorazepam monotherapy in the treatment of patients with adjustment disorders with anxiety: A double-blind controlled study in general practice. *Hum Psychopharmacol Clin Exp* 21:139–149.

Nothdurfter C, Rammes G, Baghai TC, Schüle C, Schumacher M, Papadopoulos V, Rupprecht R (2012) Translocator protein (18kDa) as a target for novel anxiolytics with a favourable side-effect profile. *J Neuroendocrinol* 24:82–92.

Nuss P, Ferreri F, Bourin M (2019) An update on the anxiolytic and neuroprotective properties of etifoxine: from brain GABA modulation to a whole-body mode of action. *Neuropsychiatr Dis Treat Volume* 15:1781–1795.

O'Beirne GB, Williams DC (1984) Binding of benzodiazepines to blood platelets from various species. *Biochem Pharmacol* 33:1568–1571.

O'Beirne GB, Williams DC (1988) The subcellular location in rat kidney of the peripheral benzodiazepine acceptor. *Eur J Biochem* 175:413–421.

O'Beirne GB, Woods MJ, Williams DC (1990) Two subcellular locations for peripheral-type benzodiazepine acceptors in rat liver. *Eur J Biochem* 188:131–138.

O'Brien CP (2005) Benzodiazepine use, abuse, and dependence. *J Clin Psychiatry* 66:28–33.

Olsen RW (2018) GABAA receptor: Positive and negative allosteric modulators. *Neuropharmacology* 136:10–22.

Olson JMM, Ciliax BJ, Mancini WR, Young AB (1988) Presence of peripheral-type benzodiazepine binding sites on human erythrocyte membranes. *Eur J Pharmacol* 152:47–53.

Papadopoulos V, Baraldi M, Guilarte TR, Knudsen TB, Lacapère JJ, Lindemann P, Norenberg MD, Nutt D, Weizman A, Zhang MR, Gavish M (2006) Translocator protein (18 kDa): New nomenclature for the peripheral-type benzodiazepine receptor based on its structure and molecular function. *Trends Pharmacol Sci* 27:402–409.

Pini S, Martini C, Abelli M, Muti M, Gesi C, Montali M, Chelli B, Lucacchini A, Cassano GB (2005) Peripheral-type benzodiazepine receptor binding sites in platelets of patients with panic disorder associated to separation anxiety symptoms. *Psychopharmacology (Berl)* 181:407–411.

Poisbeau P, Gazzo G, Calvel L (2018) Anxiolytics targeting GABA A receptors: Insights on etifoxine. *World J Biol Psychiatry* 19:S36–S45.

Ravikumar B, Crawford D, Dellovade T, Savinainen A, Graham D, Liere P, Oudinet JP, Webb M, Hering H (2016) Differential efficacy of the TSPO ligands etifoxine and XBD-173 in two rodent models of multiple sclerosis. *Neuropharmacology* 108:229–237.

Reddy SD, Younus I, Clossen BL, Reddy DS (2015) Antiseizure activity of midazolam in mice lacking δ-Subunit extrasynaptic GABAA receptors. *J Pharmacol Exp Ther* 353:517–528.

Regan JW, Yamamura HI, Yamada S, Roeske WR (1980) Renal benzodiazepine binding increases during deoxycorticosterone/salt hypertension in rats. *Eur J Pharmacol* 67:167–168.

Regan JW, Yamamura HI, Yamada S, Roeske WR (1981) High affinity renal [3H]flunitrazepam binding: Characterization, localization, and alteration in hypertension. *Life Sci* 28:991–998.

Riond J, Mattei MG, Kaghad M, Dumont X, Guillemot JC, Le Fur G, Caput D, Ferrara P (1991) Molecular cloning and chromosomal localization of a human peripheral-type benzodiazepine receptor. *Eur J Biochem* 195:305–311.

Roivainen A, Någren K, Virta J, Rinne J (2011) Human subject with unexpected biodistribution of [11C]PK11195. *Open Nucl Med J* 3:10–11.

Roncaroli F, Su Z, Herholz K, Gerhard A, Turkheimer FE (2016) TSPO expression in brain tumours: Is TSPO a target for brain tumour imaging? *Clin Transl Imaging* 4:145–156.

Rosenberg HC, Chiu TH (1985) Time course for development of benzodiazepine tolerance and physical dependence. *Neurosci Biobehav Rev* 9:123–131.

Rudolph U, Crestani F, Benke D, Brünig I, Benson JA, Fritschy JM, Martin JR, Bluethmann H, Möhler H (1999) Benzodiazepine actions mediated by specific γ-aminobutyric acid(A) receptor subtypes. *Nature* 401:796–800.

Rudolph U, Knoflach F (2011) Beyond classical benzodiazepines: Novel therapeutic potential of GABA A receptor subtypes. *Nat Rev Drug Discov* 10:685–697.

Rupprecht R et al. (2009) Translocator protein (18 kD) as target for anxiolytics without benzodiazepine-like side effects. *Science* 325:490–493.

Rupprecht R, Papadopoulos V, Rammes G, Baghai TC, Fan J, Akula N, Groyer G, Adams D, Schumacher M (2010) Translocator protein (18 kDa) (TSPO) as a therapeutic target for neurological and psychiatric disorders. *Nat Rev Drug Discov* 9:971–988.

Schippers F, Pesic M, Saunders R, Borkett K, Searle S, Webster L, Stoehr T (2020) Randomized crossover trial to compare abuse liability of intravenous remimazolam versus intravenous midazolam and placebo in recreational central nervous system depressant users. *J Clin Pharmacol* 60:1189–1197.

Schlichter R, Rybalchenko V, Poisbeau P, Verleye M, Gillardin JM (2000) Modulation of GABAergic synaptic transmission by the non-benzodiazepine anxiolytic etifoxine. *Neuropharmacology* 39:1523–1535.

Schmitz A (2016) Benzodiazepine use, misuse, and abuse: A review. *Ment Heal Clin* 6:120–126.

Schoemaker H, Boles RG, Horst WD, Yamamura HI (1983) Specific high-affinity binding sites for [3H]Ro 5-4864 in rat brain and kidney. *J Pharmacol Exp Ther* 225:61–69.

Selvaraj V, Stocco DM, Tu LN (2015) Minireview: Translocator protein (TSPO) and steroidogenesis: A reappraisal. *Mol Endocrinol* 29:490–501.

Shinotoh H, Yamasaki T, Inoue O, Itoh T, Suzuki K, Hashimoto K, Tateno Y, Ikehira H (1986) Visualization of specific binding sites of benzodiazepine in human brain. *J Nucl Med* 27:1593–1599.

Sigel E, Ernst M (2018) The benzodiazepine binding sites of GABAA receptors. *Trends Pharmacol Sci* 39:659–671.

Stein DJ (2015) Etifoxine versus alprazolam for the treatment of adjustment disorder with anxiety: A randomized controlled trial. *Adv Ther* 32:57–68.

Tournier BB, Tsartsalis S, Ceyzériat K, Garibotto V, Millet P (2020) In vivo TSPO signal and neuroinflammation in Alzheimer's disease. *Cells* 9:1941.

Trapani A, Palazzo C, De Candia M, Lasorsa FM, Trapani G (2013) Targeting of the translocator protein 18kDa (TSPO): A valuable approach for nuclear and optical imaging of activated microglia. *Bioconjug Chem* 24:1415–1428.

Tsai LH, Tsai W, Wu JY (1993) Action of myenteric GABAergic neurons in the guinea pig stomach. *Neurochem Int* 23:187–193.

Vainshtein A, Veenman L, Shterenberg A, Singh S, Masarwa A, Dutta B, Island B, Tsoglin E, Levin E, Leschiner S, Maniv I, Pe'er L, Otradnov I, Zubedat S, Aga-Mizrachi S, Weizman A, Avital A, Marek I, Gavish M (2015) Quinazoline-based tricyclic compounds that regulate programmed cell death, induce neuronal differentiation, and are curative in animal models for excitotoxicity and hereditary brain disease. *Cell Death Discov* 1:15027.

Valtier D, Malgouris C, Gilbert JC, Guicheney P, Uzan A, Gueremy C, Le Fur G, Saraux H, Meyer P (1987) Binding sites for a peripheral type benzodiazepine antagonist ([3H] PK 11195) in human iris. *Neuropharmacology* 26:549–552.

Veenman L, Gavish M (2006) The peripheral-type benzodiazepine receptor and the cardiovascular system. Implications for drug development. *Pharmacol Ther* 110:503–524.

Venneti S, Lopresti BJ, Wiley CA (2013) Molecular imaging of microglia/macrophages in the brain. *Glia* 61:10–23.

Verleye M, Heulard I, Gillardin JM (2009) The anxiolytic etifoxine protects against convulsant and anxiogenic aspects of the alcohol withdrawal syndrome in mice. *Alcohol* 43:197–206.

Wang JKT, Morgan JI, Spector S (1984) Benzodiazepines that bind at peripheral sites inhibit cell proliferation. *Proc Natl Acad Sci U S A* 81:753–756.

Werry EL, Barron ML, Kassiou M (2015) TSPO as a target for glioblastoma therapeutics. *Biochem Soc Trans* 43:531–536.

Woods MJ, Williams DC (1996) Multiple forms and locations for the peripheral-type benzodiazepine receptor. *Biochem Pharmacol* 52:1805–1814.

Section II

Naturally Occurring Benzodiazepines

6 The Search for Endogenous Benzodiazepines in Humans

Jo Ann LeQuang
NEMA Research, Inc.

CONTENTS

INTRODUCTION

The benzodiazepine (BZD) pharmaceutical products we know today were first produced in a laboratory in 1955 and came to the market in 1960 (Librium®, chlordiazepoxide) with many others introduced throughout the 1970s. They rapidly emerged as a drug class with numerous molecules with slightly different indications, such as for the treatment of anxiety, insomnia, and alcohol withdrawal syndrome. BZDs found such clinical utility that they quickly became—and remain—one of the world's most frequently prescribed drugs. In the 1970s, the discovery of endorphins, endogenous ligands acting at the opioid receptors, prompted the hypothesis that there may likewise exist endogenous BZD ligands, which were nicknamed "endozepines."[1]

After all, why would the human body have BZD receptors if it did not produce some sort of endogenous BZD? BZD receptors are part of the gamma-aminobutyric acid (GABA) system, which serves at the main inhibitory neurotransmission system of the human central nervous system. Among its many functions, GABA reduces neuronal excitability in the central nervous system and helps to regulate muscle tone. The GABA system may be further categorized by its three distinct receptor types. GABA-A is a ligand-gated ion channel. GABA-B is a metabotropic receptor and part of the class C family of G-protein-coupled receptors (GPCR), and GABA-C is a ligand-gated ion channel like GABA-A. These receptors may also include subunits. Exogenous BZDs bind to the GABA-A receptor, which may be considered the main receptor of the GABA system. GABA-A is an ionotropic receptor and ligand-gated ion channel for chloride.[2] The GABA-A receptor, which includes the BZD, when activated, selectively allows for the passage of chloride through the cell. The influx and efflux of chloride depend on the cell's resting potential millivoltage. Exogenous BZDs act at the body's BZD receptors and can potentiate the effect of GABA to increase the inflow of chloride. The well-known psychotropic effects of exogenous BZDs such as diazepam are an enhanced inhibitory tone in the brain. Exogenous BZDs along with barbiturates, ethanol, and induction anesthetics can be considered agonist ligands of the GABA-A system. These receptors and their ligands work by increasing the amplitude and/or the duration of the inhibitory postsynaptic current, which is then mediated by GABA-A receptors.

GABA-A receptors are abundant in the body. Subpopulations of these receptors in different areas of the brain exhibit varying degrees of sensitivity to GABA, steroids, and exogenous pharmaceuticals, such as BZDs.[3] The GABA-A receptor is hetero-oligomeric.[4] Exogenous BZDs are far from the only substances to show activity at the BZD receptors. In addition to agonists (BZDs), there are antagonists (flumazenil) and the so-called inverse agonists, both of which compete with agonists for access to the BZD receptors. The inverse agonists β-carboline carboxylate ethyl ester and β-carboline carboxylate methyl ester both compete for the BZD receptors, and this activity can decrease the effects of GABA, which, in turn, may promote anxiety and be proconvulsant. BZD receptor antagonists block the effects of ligands at the BZD receptors, but not in a way that affects GABA. Flumazenil and similar compounds are well-established BZD receptor antagonists and inhibit the effects of negative allosteric modulator (NAM) and positive allosteric modulator (PAM), but exert no direct action on GABA.[1] Flumazenil can induce panic attacks in patients with panic disorder, but not in healthy patients, suggesting that there is no endogenous anxiogenic (inverse agonist) ligand of the BZD receptor.[5,6] However, this might likewise argue in favor of the idea that flumazenil is an antagonist of an as-yet unidentified endogenous BZD and the panic attacks occurred when flumazenil blocked the effects of the endogenous BZD.[1]

This leaves investigators with a challenging scientific puzzle: why does the human body have BZD receptors, since BZDs are exogenous substances synthesized in a laboratory? How were these natural receptors in the body used prior to 1955? The obvious answer is that there must be some sort of endogenous BZD that acted at that receptor, and this endogenous BZD could serve as an anxiolytic and act as an anticonvulsant. Indeed, early research extracted purines and purine nucleosides from the brain that seemed to suggest the idea that, clearly, there were endogenous BZDs. However, these purines turned out to have low affinities for the BZD receptors. Neurosteroids were proposed as endogenous anxiolytics but these substances do not bind at the BZD receptors at all.[7] The discovery of endogenous opioids (endorphins) and endogenous cannabinoids (endocannabinoids) to match the body's opioid and cannabinoid receptors fueled the hypothesis that there had to be endogenous BZDs to match the body's BZD receptors.

ENDOZEPINES: FACT, FICTION, OR WISHFUL THINKING?

Considerable research has shown that there are many small molecules which possess the ability to bind to the BZD receptors. These different molecules are associated with a variety of effects. For practical purposes, they can be grouped into three main types[8]:

- PAMs, which enhance GABA functions. Traditional exogenous BZDs would be considered PAMs, along with barbiturates, ethanol, and induction anesthetics.
- NAMs such as β-carbolines, which reduce GABA functions.
- Antagonists such as flumazenil, which compete with PAMs and NAMs for the BZD receptors and, in so doing, reduce their effect.

This has given rise to the term "endozepines," which describes a class of endogenous molecules that are neither BZDs nor proteins, but which can act as positive, allosteric modulators of the GABA-A receptors and, in that way, may have a similar activity to exogenous BZDs.[9] Endozepines are PAMs, but they also appear to have multiple biological functions. Such endogenous peptides have been identified in a number of vertebrate species, where they are widely distributed throughout the body in both peripheral and central nervous systems.[10] Endozepines have effects not only at the central BZD receptors, but also at peripheral and metabotropic receptors. Furthermore, in murine studies, the injection of endozepines into the intracerebroventricle confers effects that are in diametric opposition to those of diazepam: endozepines injected into the brain are anxiogenic, encourage conflict behaviors, and may increase depression. Although the reasons for this remain unclear, endozepines similarly injected into the brains of pigs had no effect. This suggests that these endozepines are not endogenous BZDs, but rather inverse agonists at the BZD receptors.[10] It should be noted that while endozepines are similar to exogenous BZDs, they are not halogenated.[11]

While PAM endozepines have not yet been identified in humans, it is not for lack of study. The search for endozepines started in the 1980s with several candidate molecules proposed, including peptides, fatty acid derivatives, and small organic molecules (purines). Naturally occurring BZDs have been identified in animals as molecules with a benzene ring fused to a diazepine ring.[12,13] Some of the first laboratory efforts to find endozepines used radioligand-binding assays: with agonists [^{3}H]FNZ and [^{3}H]diazepam, peripherals [^{3}H]Ro 5-4864, the antagonist [^{3}H]Ro 15-1788, and the inverse agonist [^{3}H]βCCE.[7] In immunocytochemical experiments on the brains of rats, it was found that mAb 21-7F9 exhibited immunoreactivity that could be blocked using 10^{-7M} diazepam and other BZDs, but that β-carbolines and Ro15-1788 did not have this same effect.[7] Researchers then worked on immunoreactive substances in bovine brains by immobilizing mAB 21-79 on Affi-gel and passing brain extracts through an immunoaffinity column. Endogenous material was obtained and eluted via decreased pH. Thus purified, it was found that this endogenous substance could bind to neuronal GABA-A receptors, specifically the BZD receptors, for which they showed great affinity. In further studies, other agents were employed to compete with this endogenous substance for the BZD receptors, confirming the great affinity of the endogenous substance. In these studies, it was also found that the GABA system potentiated the inhibition of [^{3}H]Ro15-1788 from binding to the BZD receptors. This suggested that this endogenous substance was a BZD receptor agonist. In a series of laboratory steps, this substance was further purified, whereupon it was reported to be *N*-desmethyldiazepam, present naturally in both the murine and bovine brain tissues.[7,14,15] *N*-Desmethyldizepam is a metabolite formed in the body after metabolism of exogenous BZDs, such as diazepam and others. This metabolite can be used in and of itself as an exogenous pharmacological BZD. It is slowly cleared from the body, and as it is oxidized, it is converted to oxazepam, which is then conjugated to glucuronic acid, thus resulting in glucuronide that can be excreted via the kidneys.[7]

FUTURE DIRECTIONS

The search for the elusive endogenous BZD has resulted in some important new findings. From these studies, for example, it has been established that the octadecaneuropeptide (ODN) (DB133-50) exhibits activity at the central BZD receptors, where it has metabotropic activities such as enhancing pro-conflict behavior, inducing anxiety, decreasing consumption of food and water, and inhibiting phenobarbital-induced sleep.[16] As such, it would be a NAM drug and not the missing endogenous PAM. But ODN is of great interest in medicine, as it regulates glial cell and neurons and can stimulate astrocyte proliferation. It has been suggested that ODN exerts a protective effect on neurons and astrocytes against cell death induced by oxidative stress. Overall, ODN has an effect on the body's natural apoptotic cascade, which makes it an interesting potential drug target, for example, for treating cerebrovascular accident and neurodegeneration.[16]

Endozepine-4 is an endogenous compound and ligand for the BZD receptor, and is one that has garnered considerable attention as a possible endogenous BZD. Recent research suggests endozepine-4 plays a role in the pathogenesis of hepatic encephalopathy, a neuropsychiatric condition in which cerebral function deteriorates with liver failure. Ammonia has long been recognized as playing a crucial, but largely unelucidated role in the etiology of hepatic encephalopathy. Patients with hepatic encephalopathy typically exhibit cerebral swelling as a result of the buildup of extracellular fluids. These fluids are the accumulated toxic products not otherwise metabolized by the liver combined with ammonia. The latter is a potent neurotoxin that damages brain metabolism and can destroy astrocytes. The neurotoxicity associated with ammonia is related to glutamate, which can cause changes to the GABA-A receptor system. In animal models and human patients, BZD receptor antagonists have been shown to be able to revert the symptoms of hepatic encephalopathy.[9] In a clinical study of 20 patients with hepatic coma, 20 patients with minimal hepatic encephalopathy, and 20 controls, all patients were analyzed for the presence of endoazepine-4 as well as other measures of hepatic disease. Patients in hepatic coma had significantly increased levels of endozepine-4 ($p < 0.001$) compared to those with mild disease and controls. Likewise, ammonia levels were highest in comatose patients, and ammonia levels were correlated with endodzepine-4 levels ($p < 0.001$).[9] Thus, endozepines were shown to be highly correlated with the degree of hepatic encephalopathy. The high affinity that endozepines exhibit for the GABA-A receptors may be the reason that these endogenous substances can induce coma.[9]

Endozepine-4 has also been implicated in a controversial condition termed "endozepine stupor," in which patients exhibit recurrent episodes of stupor combined with elevated serum endozepine-4 levels.[10,17] Endozepine stupor is sufficiently rare that some have suspected it is misdiagnosed as the result of malignant hypersomnia or covert drug ingestion.[10,18] Patients surreptitiously taking exogenous BZDs can be screened for those drugs, but endozepines will not appear on a drug assay.[10]

An acyl-CoA-binding protein (10 kDa) has been discovered in the search for endogenous BZDs. This protein acts as a diazepam-binding inhibitor (DBI).[19] The elimination of the gene that encodes the acyl-CoA-binding protein is associated with negative allosteric modulation of the GABA system. Thus, the DBI protein may be considered the endogenous NAM. The DBI is of interest to oncology, in that the acyl-CoA-binding protein appears to drive the growth of glioblastoma by promoting the accumulation of mitochondrial long fatty acyl-CoA accumulation and β-oxidation. It should be noted that in glioblastoma, altered lipid metabolism is often present, but it is not clear what its role is in tumorigenesis.[20]

STILL LOOKING FOR THE ENDOGENOUS PAM

A dietary hypothesis for the presence of endogenous BZDs has been around for decades. While considered by some to be implausible, the dietary theory notes that naturally occurring BZDs are present in certain foods, so perhaps endogenous BZDs in the body have entered via food sources. Indeed, certain BZDs found in the body have never been synthesized in a laboratory, such as isodiazepam and deschlorodiazepam.[21] Substances similar to BZDs have been found in potatoes, cereals, and certain vegetables.[21] BZD-like molecules have been identified in human milk in women not taking exogenous BZDs.[22] It has also been theorized that at least some endogenous BZDs are synthesized by the neurological system. For instance, mammals might be able to synthesize BZDs using pathways similar to those used by *Penicillium cyclopium*, which rely on anthranilic acid and phenylalanine as precursors.[23] Plant sources for substances like opiates and tetrahydrocannabinol that bind to opioid and cannabinoid receptors seem to support this dietary theory, although it does not preclude the hypothesis that there is (also) an endogenous PAM BZD produced naturally in the body.

CONCLUSION

PAMs, such as exogenous BZDs like lorazepam and diazepam, target the GABA-A receptor, and while much is known about the GABA system, the endogenous PAM remains elusive. In fact, no plausible pathway for its production within the body has been identified. This argues in favor of the food hypothesis, that is, endozepines are likely ingested from outside sources rather than synthesized internally.[11] It should be noted that these two theories are not mutually exclusive; that is, the body may produce an endogenous PAM while still obtaining some BZDs through food. Numerous endogenous substances that affect the GABA system have been identified, and it is likely that at some point in the future, a clear endogenous PAM or a pathway for the creation of endogenous BZDs will be discovered.

REFERENCES

1. Farzampour Z, Reimer RJ, Huguenard J. Endozepines. *Adv Pharmacol*. 2015;72:147–164.
2. Schofield PR, Darlison MG, Fujita N, et al. Sequence and functional expression of the GABA A receptor shows a ligand-gated receptor super-family. *Nature*. 1987;328(6127): 221–227.

3. Olsen RW, Tobin AJ. Molecular biology of GABAA receptors. *FASEB J.* 1990;4(5): 1469–1480.
4. Pritchett DB, Sontheimer H, Shivers BD, et al. Importance of a novel GABAA receptor subunit for benzodiazepine pharmacology. *Nature.* 1989;338(6216):582–585.
5. Nutt DJ, Glue P, Lawson C, Wilson S. Flumazenil provocation of panic attacks. Evidence for altered benzodiazepine receptor sensitivity in panic disorder. *Arch Gen Psychiatry.* 1990;47(10):917–925.
6. Potokar J, Lawson C, Wilson S, Nutt D. Behavioral, neuroendocrine, and cardiovascular response to flumazenil: no evidence for an altered benzodiazepine receptor sensitivity in panic disorder. *Biol Psychiatry.* 1999;46(12):1709–1711.
7. De Blas A. Benzodiazepines and benzodiazepine-like molecules are present in brain. In: Izquierdo I, Medina J, eds. *Naturally Occurring Benzodiazepins: Structure, Distribution and Function.* New York: Ellis Horwood; 1993:1–27.
8. Harward SC, McNamara JO. In search of the ever-elusive positive endozepine. *Neuron.* 2013;78(6):951–952.
9. Malaguarnera G, Vacante M, Drago F, et al. Endozepine-4 levels are increased in hepatic coma. *World J Gastroenterol.* 2015;21(30):9103–9110.
10. Granot R, Berkovic SF, Patterson S, Hopwood M, Drummer OH, Mackenzie R. Endozepine stupor: disease or deception? A critical review. *Sleep.* 2004;27(8):1597–1599.
11. Baraldi M, Avallone R, Corsi L, Venturini I, Baraldi C, Zeneroli ML. Endogenous benzodiazepines. *Therapie.* 2000;55(1):143–146.
12. Medina JH, Peña C, Novas ML, Paladini AC, De Robertis E. Acute stress induces an increase in rat cerebral cortex levels of n-butyl-?-carboline-3-carboxylate, an endogenous benzodiazepine binding inhibitor. *Neurochem Int.* 1987;11(3):255–259.
13. Medina JH, Peña C, Piva M, Paladini AC, De Robertis E. Presence of benzodiazepine-like molecules in mammalian brain and milk. *Biochem Biophys Res Commun.* 1988; 152(2):534–539.
14. Ruano D, Araujo F, Machado A, de Blas AL, Vitorica J. Molecular characterization of type I GABAA receptor complex from rat cerebral cortex and hippocampus. *Brain Res Mol Brain Res.* 1994;25(3–4):225–233.
15. Robertis ED, Peña C, Paladini AC, Medina JH. New developments on the search for the endogenous ligand(s) of central benzodiazepine receptors. *Neurochem Int.* 1988;13(1):1–11.
16. Masmoudi-Kouki O, Hamdi Y, Ghouili I, et al. Neuroprotection with the endozepine octadecaneuropeptide, ODN. *Curr Pharm Des.* 2018;24(33):3918–3925.
17. Lugaresi E, Montagna P, Tinuper P, et al. Endozepine stupor. Recurring stupor linked to endozepine-4 accumulation. *Brain.* 1998;121(Pt 1):127–133.
18. Lugaresi E, Montagna P, Tinuper P, Plazzi G, Gallassi R. Suspected covert lorazepam administration misdiagnosed as recurrent endozepine stupor. *Brain.* 1998; 121(Pt 11):2201.
19. Knudsen J. Acyl-CoA-binding and transport, an alternative function for diazepam binding inhibitor (DBI), which is identical with acyl-CoA-binding protein. *Neuropharmacology.* 1991;30(12b):1405–1410.
20. Bi J, Mischel PS. Acyl-CoA-binding protein fuels gliomagenesis. *Cell Metab.* 2019;30(2):229–230.
21. Wildmann J. Heterocycles as physiological ligands for the benzodiazepine receptor and for other binding sites. *Pharmacol Res.* 1989;21(6):673–682.
22. Medina J, Peña C, Piva M, Rubio M, Wikinski S, Paladini A. Benzodiazepine-like molecules in human milk. *Lancet.* 1990;336(8727):1379.
23. Luckner M, Nover L. Expression of secondary metabolism. An aspect of cell specialization of microorganisms, higher plants, and animals. *Mol Biol Biochem Biophys.* 1977;23:3–102.

7 Endozepines in Insect Development and Metamorphosis

Oné R. Pagán
West Chester University

CONTENTS

EVERYTHING BEGINS WITH BINDING

The history of neuropharmacology is rich in examples of receptor systems discovered by observing the actions of exogenous chemicals in living organisms (Feng et al., 2017; Robinson, 2018). In humans, many of these examples involve the historical consumption of plants producing psychoactive compounds that eventually became abused drugs. Three of the best-known examples of such drugs are cocaine, nicotine, and opiates.

The consumption of coca leaves by South American natives led to the emergence of the local anesthetic field through the identification of cocaine, and its subsequent identification as a sodium channel inhibitor and as an inhibitor of monoamine transporter function (Biondich and Joslin, 2016; Ruetsch et al., 2001). Tobacco consumption and the identification of its main addictive chemical, nicotine, defined a whole cholinergic receptor class, the nicotinic acetylcholine receptor, involved in multiple neurological and neuromuscular processes and pathologies (Changeux, 2020; Fletcher, 1941). In contrast with the examples of cocaine and nicotine, the identification of morphine as the main psychoactive compound in poppy plants (Evans, 2004; Presley and Lindsley, 2018) led not only to the identification of its receptor targets (the opiate receptors), but also to the discovery of the receptors' previously unknown endogenous peptide ligands: the endorphins (Snyder, 2004; Fricker et al., 2020; Hughes et al., 1975; Papini, 2018).

The earliest formal articulation of the receptor concept came from Paul Ehrlich, a Nobel laureate who is generally recognized as the father of modern chemotherapy (Bosch and Rosich, 2008). Ehrlich's famous maxim, *Corpora non agunt nisi fixata* (*Physical entities do not interact unless they are bound to each other*; Gesztelyi et al., 2012; Pelner, 1972), is one of my favorite scientific quotes. In fact, in one of my works, I translated it as *Everything begins with binding* (Pagán, 2005). The discovery of endozepines is a perfect example to illustrate how receptor binding techniques are one of the main tools that we have in order to achieve pharmacological discoveries.

A BRIEF HISTORY OF THE IDENTIFICATION OF THE BENZODIAZEPINE SITES IN MAMMALS

Benzodiazepines have been in medical use since the 1960s to treat conditions such as anxiety (Tallman et al., 1980). However, up to the early 1970s, their precise mechanism of action to induce their antianxiety effects was unknown. We know now that the mechanism of benzodiazepines for their interaction with the GABA-A receptor relates to their positive allosteric modulation of GABA-A receptors in the central nervous system of vertebrates (Costa, 1998; Guidotti et al., 1978; Möhler, 2015; Olsen et al., 2004). In addition to their effects on GABA-A receptors, benzodiazepines also interact with a series of peripheral-type receptors, termed "peripheral benzodiazepine receptors." These are translocator proteins, which are cholesterol transporters mainly located in the outer mitochondrial membrane of neuronal and non-neuronal tissues, as well as in microorganisms (Papadopoulos, 1993, 2003, 2014; Papadopoulos et al., 1991). This molecule is evolutionarily conserved, with versions of it found from prokaryotes to mammals (Bonsack et al., 2018; Fan et al., 2012). These translocator proteins are thought to be involved in a variety of normal and pathological processes (Lee et al., 2020; Papadopoulos, 2003).

Benzodiazepines are not naturally occurring compounds. In fact, these compounds were developed in a serendipitous way. In the early 1950s, a research group based at Hoffmann-La Roche in New Jersey was tasked with the discovery of possible antianxiety compounds. Instead of trying to modify the existing drugs, they decided to seek for new classes of compounds, hoping to get a few candidates with the desired pharmacological effect. For the first couple of years, their efforts were not fruitful, and the focus of the group veered off in the direction of the search for novel antibiotics. Their luck changed in 1957 in the form of a much-needed "...major spring cleaning" (Sternbach, 1979), due to the messiness of the laboratory. A laboratory member, Dr. Earl Reeder, found two vials with purified compound in its acid and base forms, which were "...made sometime in 1955." The group director, Dr. Leo Sternbach, decided to send the samples for behavioral testing. Their pharmacologist, Dr. Lowell Randall, called Dr. Sternbach with promising results. The candidate compound showed activity in models of anxiety via a series of *in vivo* assays, and therefore, the group abandoned their antibiotic research and returned to their previous efforts to find new antianxiety drugs. They tinkered with this initial compound, eventually coming up with the first commercially available benzodiazepine, Librium, colloquially known as chlordiazepoxide, released in 1960. Remarkably, the time elapsed between its synthesis and testing and its commercial launch was only two and a half years (Sternbach, 1979). Valium (diazepam), which is arguably the most famous benzodiazepine, was approved for clinical use only 2 years after the approval of Librium. Since then, many benzodiazepine derivatives have been synthesized, and some 15 of those are currently approved for clinical use (McIntosh et al, 2011), with a worldwide average of 100 million prescriptions per year as of 2007 (Griffin et al., 2013). These drugs are popular in part because of their anxiolytic properties; in addition, they are also sedatives, anticonvulsants, and muscle relaxants (Baldwin et al., 2013; Griffin et al., 2013).

The obvious targets of the benzodiazepines were located in the CNS; therefore, researchers investigated the location of these putative sites using radioligand binding techniques in rat brain membrane preparations. Their results, added to the lessons learned from the opiate receptor story, led investigators to propose the existence of endogenous molecules with benzodiazepine-like properties.

In 1977, two research groups independently demonstrated the high-affinity binding of radiolabeled diazepam to rat brain membranes (Möhler and Okada, 1977; Squires and Braestrup, 1977). The authors of one of the papers explicitly stated, "We therefore conclude that the binding site for [^{3}H]-diazepam is physiologically relevant and that an unknown endogenous transmitter acting on these receptors may exist" (Squires and Braestrup, 1977). This was the first allusion to a putative naturally occurring benzodiazepine binding site. In an editorial commenting on the aforementioned paper, the late pharmacologist Leslie Iversen stated, "By analogy with the discovery of the opiate receptor, the benzodiazepine receptor assay may even offer a simple means for the discovery of the naturally occurring ligand for such receptors, the brain's own anti-anxiety substance? Further developments in this field will certainly be awaited—anxiously." (Iversen, 1977). For an overview of additional historical aspects of the discovery of the benzodiazepine binding sites, please see Farzampour et al. (2015), Costa and Guidotti (1985, 1991), and Möhler (2015).

THE DISCOVERY OF ENDOZEPINES

Prompted by the work of Möhler and Okada and Squires and Braestrup, many research groups engaged in an intense effort to identify naturally occurring molecules displaying benzodiazepine-like effects (endozepines). Endogenous candidate molecules included various small proteins, lipids, peptides, and small organic molecules (Farzampour et al., 2015; Tonon et al., 2020; Wick, 2013). Most of these ligands, although undoubtedly binding to benzodiazepine sites, did not display relevant physiological affinity (Tonon et al., 2020).

In 1978, two independent groups reported the presence of an endogenous peptide that displaced the binding of [^{3}H]diazepam in rat brain membranes (Guidotti et al., 1978; Marangos et al., 1978; Toffano et al., 1978). Later on, this peptide was characterized biochemically and physiologically, and was eventually called diazepam-binding inhibitor (DBI; Guidotti et al., 1983). DBIs are highly conserved proteins of 82–92 amino acids long (Faergeman et al., 2007; Mocchetti et al., 1986). The bovine version was the one originally termed "endozepine" (Shoyab et al., 1986). In 1989, a protein identified as a lipid carrier with a significant binding affinity to acyl-coenzyme-A esters (Knudsen et al., 1989) was isolated from the rat and bovine liver. This protein was termed "acyl-CoA-binding protein" (ACBP). Surprisingly, the primary sequence of the ACBP turned out to be identical to the primary sequence of the DBI (Knudsen et al., 1993). In many subsequent publications, the protein is designated as DBI/ACBP (Tonon et al., 2020). Genetic analysis across phylogenetic lines made apparent the fact that the DBI/ACBP plays essential roles in cells, as there is evidence that this protein has been in biological systems for almost 2 billion years (Burton et al., 2005; Faergeman et al., 2007). Depending on the species, up to 56 of the 86 residues of the protein are identical in nature and position in organisms from yeast to mammals (Faergeman et al., 2015; Tonon et al., 2020). This high degree of conservation of the ACBP protein eukaryotes and the ubiquitous characteristic of its tissue distribution argues in favor of a basal function shared by virtually all cell types. Even though the precise fundamental biological function of this protein remains elusive at the moment, the best available evidence seems to indicate that it is required for sphingolipid synthesis and fatty acid chain elongation, among other related cell functions (Faergeman et al., 2007). All this information strongly hints at the importance of the DBI/ACBP protein, likely in a variety of biological processes. The various functions of this protein in a variety of organisms are explored elsewhere in this volume. Thus, I decided to touch upon a relative obscure function of the DBI/ACBP protein: its role in invertebrate development and metamorphosis.

DBI/ACBP IN INSECT REPRODUCTION

The presence and function of the DBI/ACBP proteins have been demonstrated in various insect species. *Rhodnius prolixus* is a type of kissing bug whose claim to fame is being one of the vectors of Chagas disease. The *R. prolixus* genome includes five ACBP genes, with the expected functions related to lipid physiology. When this protein was knocked down through RNAi, triacylglycerol accumulated abnormally in the insect hind midgut. Also, this knockdown caused a significant reduction in the number of deposited eggs, suggesting an ACBP role in reproduction. In contrast, the normal reserves of triacylglycerol were depleted in *R. prolixus* flight muscles, and cholesteryl ester fatty acids abnormally accumulated in the fat body (Majerowicz et al., 2016). Interestingly, the neurotransmitter serotonin regulates the expression of ACBP in the gut of *R. prolixus* (Alves-Bezerra et al., 2010).

Another aspect of the expression of ACBP related to insect reproduction is the example of the pheromone glands of the silkworm, *Bombyx mori*. In these insects, ACBP is expressed mainly in their pheromone glands, likely due to an effect on the production of these compounds, traced by the ACBP-induced activation of their "pheromone biosynthesis activating neuropeptide" (Matsumoto et al., 2001). One of the factors that influence the biosynthesis of ACBP in *B. mori* is the amino acid-derivative beta-D-glucosyl-O-L-tyrosine. This was the first demonstration of humoral factors influencing the transcription of ACBP genes (Ohnishi et al., 2005).

DBI/ACBP IN INSECT METAMORPHOSIS AND DEVELOPMENT

Development is one of the most interesting and mysterious processes in all of biology, and metamorphosis is a fascinating subset of developmental biology. In fact, one can think of all of developmental processes as metamorphosis-like events, as in most cases of development, the organism's adult behavior, morphology, and physiology might be drastically different from the same parameters at the beginning of the organism's life. As you may imagine, arguably one of the most dramatic examples of metamorphosis is insect metamorphosis. For example, a caterpillar looks nothing like a butterfly; upon seeing them for the first time, an early scientist unaware of metamorphosis would surely classify those two entities as separate species. As remarkable as this process is at a first approximation, it becomes even more amazing if we realize that the transformation of the embryonic form of an insect into an adult entails an almost complete reorganization of bodily tissues. And yet, amazingly, and against what anyone would expect, it is evident that some information pertaining the embryonic form is conserved. Case in point: insect memories survive the central nervous system rearrangement and remodeling inherent to metamorphosis (Blackiston et al., 2015). Many of the most extreme examples of metamorphosis appear in the Ecdysozoans, a superphylum comprising arthropods, nematodes, and related phyla.

The defining characteristic of Ecdysozoans is their cuticle, which is a hard, yet flexible exoskeleton that has a protective function against environmental insults or predators (Pan et al., 2020). Due to the nature of their exoskeleton, as part of their growth processes, all members of this superphylum must molt and discard their cuticle as they develop and grow. As expected, molting and ecdysis (cuticle shedding) also play an integral process in metamorphosis (Truman, 2019). A type of molecules with an essential role in metamorphosis are the Ecdysteroids; as indicated by their name, these are Ecdysozoans-specific steroid hormones that play a central role in developmental processes like molting (particularly ecdysis) and development (and of course, metamorphosis), as well as having a relatively minor role in reproduction. The synthesis of ecdysteroids is termed ecdysteroidogenesis, a process controlled by multiple factors (Marchal et al., 2010).

The cotton bollworm (*Helicoverpa armigera*) is a small moth that is considered an agricultural pest for a variety of crops, predominantly cotton. In this insect, the expression of ACBP is closely linked with hormone levels at ecdysis displaying probable upregulation at various stages of molting (Wang et al., 2008). Interestingly, the expression of ACBP mRNA in the brain of nondiapause (actively molting) insects was significantly higher than in brains of diapausing (undergoing arrested development) insects, and in fact, ACBP levels fluctuated in response to the specific developmental state in this organism, strongly suggesting a role in tissue remodeling during metamorphosis (Liu et al., 2005).

ACBP proteins also seem to have a prominent role in ecdysteroidogenesis. In the tobacco hornworm (*Manduca sexta*), high levels of ACBP expression correlate with the rate of ecdysteroidogenesis, strongly suggesting a role in metamorphosis (Snyder and Antwerpen, 1997). Furthermore, in the aforementioned work, this relationship was pharmacologically verified by treating the insects with a benzodiazepine analog (FGIN-1–27), an anxiolytic drug that is a selective agonist at mammalian peripheral benzodiazepine receptors. Exposure to FGIN-1–27 increased ecdysteroidogenesis, while exposure to anti-ACBP antibodies inhibited ecdysteroidogenesis.

PERSPECTIVES

To say that endozepines and their associated receptors are interesting molecules is an understatement. Their ubiquity in biological systems, their multiple functions, as well as their phylogenetic conservation in plants, animals, and microorganisms hint at a profound significance of this pharmacological system with multiple fundamental scientific implications and possible applications to the biomedical sciences. In a somewhat related note, I could not help but to think of the path of discovery of the endozepines. When I think about the search for novel anxiolytics, the identification of their binding sites in mammalian brains, the generation of synthetic compounds targeted to the GABA-A receptor system that led to medications, the notion of a natural benzodiazepine, the discovery of such compounds, and their possible significance to biomedicine, I could not help but thinking:

Where are the brain's natural barbiturates?

ACKNOWLEDGMENTS

I am grateful for the support from the Department of Biology and the College of Science and Mathematics, West Chester University.

CONFLICT OF INTEREST STATEMENT

The author declares no conflict of interest.

REFERENCES

Alves-Bezerra M, Majerowicz D, Grillo LA, Tremonte H, Almeida CB, Braz GR, Sola-Penna M, Paiva-Silva GO, Gondim KC. Serotonin regulates an acyl-CoA-binding protein (ACBP) gene expression in the midgut of *Rhodnius prolixus*. *Insect Biochem Mol Biol.* 2010;40(2):119–25.

Baldwin DS, Aitchison K, Bateson A, Curran HV, Davies S, Leonard B, Nutt DJ, Stephens DN, Wilson S. Benzodiazepines: risks and benefits. A reconsideration. *J Psychopharmacol.* 2013 Nov;27(11):967–71.

Biondich AS, Joslin JD. Coca: The history and medical significance of an ancient Andean tradition. *Emerg Med Int.* 2016;2016:4048764.

Blackiston DJ, Shomrat T, Levin M. The stability of memories during brain remodeling: A perspective. *Commun Integr Biol.* 2015;8(5):e1073424.

Bonsack F, Sukumari-Ramesh S. TSPO: An evolutionarily conserved protein with elusive functions. *Int J Mol Sci.* 2018;19(6):1694.

Bosch F, Rosich L. The contributions of Paul Ehrlich to pharmacology: A tribute on the occasion of the centenary of his Nobel Prize. *Pharmacology.* 2008;82(3):171–9.

Burton M, Rose TM, Faergeman NJ, Knudsen J. Evolution of the acyl-CoA binding protein (ACBP). *Biochem J.* 2005;392(Pt 2):299–307.

Changeux JP. Discovery of the first neurotransmitter receptor: The acetylcholine nicotinic receptor. *Biomolecules.* 2020;10(4):547.

Costa E. From GABAA receptor diversity emerges a unified vision of GABAergic inhibition. *Annu Rev Pharmacol Toxicol.* 1998;38:321–50.

Costa E, Guidotti A. Endogenous ligands for benzodiazepine recognition sites. *Biochem Pharmacol.* 1985;34(19):3399–403.

Costa E, Guidotti A. Diazepam binding inhibitor (DBI): A peptide with multiple biological actions. *Life Sci.* 1991;49(5):325–44.

Evans CJ. Secrets of the opium poppy revealed. *Neuropharmacology.* 2004;47 Suppl 1:293–9.

Faergeman NJ, Wadum M, Feddersen S, Burton M, Kragelund BB, Knudsen J. Acyl-CoA binding proteins; structural and functional conservation over 2000 MYA. *Mol Cell Biochem.* 2007;299(1–2):55–65.

Fan J, Lindemann P, Feuilloley MG, Papadopoulos V. Structural and functional evolution of the translocator protein (18kDa). *Curr Mol Med.* 2012;12:369–86.

Farzampour Z, Reimer RJ, Huguenard J. Endozepines. *Adv Pharmacol.* 2015;72:147–64.

Feng LY, Battulga A, Han E, Chung H, Li JH. New psychoactive substances of natural origin: A brief review. *J Food Drug Anal.* 2017;25(3):461–71.

Fletcher HG. The history of nicotine. *J Chem Educ.* 1941;18:7, 303.

Fricker LD, Margolis EB, Gomes I, Devi LA. Five decades of research on opioid peptides: Current knowledge and unanswered questions. *Mol Pharmacol.* 2020;98(2):96–108.

Gesztelyi R, Zsuga J, Kemeny-Beke A, Varga B, Juhasz B, Tosaki A. The Hill equation and the origin of quantitative pharmacology. *Arch Hist Exact Sci.* 2012;66(4):445–54.

Griffin, CE, III, Kaye, AM, Bueno, FR, & Kaye, AD. Benzodiazepine phar-macology and central nervous system-mediated effects. *Ochsner J*, 2031;13, 214–23.

Guidotti A, Forchetti CM, Corda MG, Konkel D, Bennett CD, Costa E. Isolation, characterization, and purification to homogeneity of an endogenous polypeptide with agonistic action on benzodiazepine receptors. *Proc Natl Acad Sci U S A.* 1983;80:3531–5.

Guidotti A, Toffano G, Costa E. An endogenous protein modulates the affinity of GABA and benzodiazepine receptors in rat brain. *Nature.* 1978;275:553–5.

Hughes J, Smith TW, Kosterlitz HW, Fothergill LA, Morgan BA, Morris HR. Identification of two related pentapeptides from the brain with potent opiate agonist activity. *Nature.* 1975;258(5536):577–80.

Iversen L. Anti-anxiety receptors in the brain? *Nature.* 1977;266:678.

Knudsen J, Højrup P, Hansen HO, Hansen HF, Roepstorff P. Acyl-CoA-binding protein in the rat. Purification, binding characteristics, tissue concentrations and amino acid sequence. *Biochem J.* 1989;262:513–9.

Knudsen J, Mandrup S, Rasmussen JT, Andreasen PH, Poulsen F, Kristiansen K. The function of acyl-CoA-binding protein (ACBP)/diazepam binding inhibitor (DBI). *Mol Cell Biochem.* 1993;123:129–38.

Lee Y, Park Y, Nam H, Lee JW, Yu SW. Translocator protein (TSPO): The new story of the old protein in neuroinflammation. *BMB Rep.* 2020;53(1):20–7.

Liu M, Zhang TY, Xu WH. A cDNA encoding diazepam-binding inhibitor/acyl-CoA-binding protein in *Helicoverpa armigera*: Molecular characterization and expression analysis associated with pupal diapause. *Comp Biochem Physiol C Toxicol Pharmacol.* 2005; 141(2):168–76.

Majerowicz D, Hannibal-Bach HK, Castro RSC, Bozaquel-Morais BL, Alves-Bezerra M, Grillo LAM, Masuda CA, Færgeman NJ, Knudsen J, Gondim KC. The ACBP gene family in *Rhodnius prolixus*: Expression, characterization and function of RpACBP-1. *Insect Biochem Mol Biol.* 2016;72:41–52.

Marangos PJ, Paul SM, Greenlaw P, Goodwin FK, Skolnick P. Demon-stration of an endogenous, competitive inhibitor(s) of [3H]diazepam binding inbovine brain. *Life Sci.* 1978; 22:1893–900.

Marchal E, Vandersmissen HP, Badisco L, Van de Velde S, Verlinden H, Iga M, Van Wielendaele P, Huybrechts R, Simonet G, Smagghe G, Vanden Broeck J. Control of ecdysteroidogenesis in prothoracic glands of insects: A review. *Peptides.* 2010;31(3):506–19.

Matsumoto S, Yoshiga T, Yokoyama N, Iwanaga M, Koshiba S, Kigawa T, Hirota H, Yokoyama S, Okano K, Mita K, Shimada T, Tatsuki S. Characterization of acyl-CoA-binding protein (ACBP) in the pheromone gland of the silkworm, *Bombyx mori. Insect Biochem Mol Biol.* 2001;31(6–7):603–9.

McIntosh B, Clark M, Spry C. Benzodiazepines in Older Adults: A Review of Clinical Effectiveness, Cost-Effectiveness, and Guidelines [Internet]. Ottawa: Canadian Agency for Drugs and Technologies in Health; 2011 (Rapid Response Report: Peer-Reviewed Summary with Critical Appraisal).

Mocchetti I, Einstein R, Brosius J. Putative diazepam binding inhibitorpeptide: CDNA clones from rat. *Proc Natl Acad Sci U S A.* 1986;83:7221–5.

Möhler, H. The legacy of the benzodiazepine receptor: From flumazenilto enhancing cognition in Down syndrome and social interaction in autism. *Adv Pharmacol.* 2015;72:1–36.

Möhler H, Okada T. Benzodiazepine receptor: Demonstration in the central nervous system. *Science.* 1977; 198(4319):849–51.

Ohnishi A, Koshino H, Takahashi S, Esumi Y, Matsumoto S. Isolation and characterization of a humoral factor that stimulates transcription of the acyl-CoA-binding protein in the pheromone gland of the silkmoth, *Bombyx mori. J Biol Chem.* 2005;280(6):4111–6.

Olsen RW, Chang CS, Li G, Hanchar HJ, Wallner M. Fishing for allosteric sites on GABA(A) receptors. *Biochem Pharmacol.* 2004 Oct 15;68(8):1675–84.

Pagán OR. 2005. Synthetic Local Anesthetics as Alleviators of Cocaine Inhibition of the Human Dopamine Transporter. PhD dissertation. Cornell University, Ithaca, NY.

Pan X, Connacher RP, O'Connor MB. Control of the insect metamorphic transition by ecdysterioid production and secretion. *Curr Opin Insect Sci.* 2020;43:11–20.

Papadopoulos V. Peripheral-type benzodiazepine/diazepam binding inhibitor receptor: Biological role in steroidogenic cell function. *Endocrine Rev.* 1993;14:222–40.

Papadopoulos V. Peripheral benzodiazepine receptor: Structure and function in health and disease. *Ann Pharm Fr.* 2003;61(1):30–50.

Papadopoulos V. On the role of the translocator protein (18-kDa) TSPO in steroid hormone biosynthesis. *Endocrinology.* 2014;155:15–20.

Papadopoulos V, Berkovich A, Krueger KE, Costa E, Guidotti A. Diazepam binding inhibitor and its processing products stimulate mitochondrial steroid biosynthesis via an interaction with mitochondrial benzodiazepine receptors. *Endocrinology.* 1991;129:1481–8.

Papini AM. From morphine to endogenous opioid peptides, e.g., endorphins: The endless quest for the perfect painkiller. *Substantia.* 2018;2(2):81–91.

Pelner L. Corpora non agunt nisi fixata. Maxim behind all of Ehrlich's great discoveries. *NY State J Med.* 1972;72:620–4.

Presley CC, Lindsley CW. DARK classics in chemical neuroscience: Opium, a historical perspective. *ACS Chem Neurosci.* 2018;9(10):2503–518.

Robinson E. Psychopharmacology: From serendipitous discoveries to rationale design, but what next? *Brain Neurosci Adv.* 2018;2:2398212818812629.

Ruetsch YA, Böni T, Borgeat A. From cocaine to ropivacaine: The history of local anesthetic drugs. *Curr Top Med Chem.* 2001;1(3):175–82.

Shoyab M, Gentry LE, Marquardt H, Todaro GJ. Isolation and characterization of a putative endogenous benzodiazepineoid (endozepine) from bovine and human brain. *J Biol Chem.* 1986;261:11968–73.

Snyder SH. Opiate receptors and beyond: 30 years of neural signaling research. *Neuropharmacology.* 2004;47 Suppl 1:274–85.

Snyder MJ, Antwerpen RV. Cellular distribution, levels, and function of the diazepam-binding inhibitor/acyl-CoA-binding protein in last instar *Manduca sexta* midgut. *Cell Tissue Res.* 1997;288(1):177–84.Squires RF, Braestrup A. Benzodiazepine receptors in rat brain. *Nature.* 1977;266:732–4.

Tallman JF, Paul SM, Skolnick P, Gallager DW. Receptors for the age of anxiety: pharmacology of benzodiazepines. *Science* 1980;207:274–281.

Sternbach LH. The benzodiazepine story. *Prog Drug Res.* 1978;22:229–266.

Toffano G, Guidotti A, Costa E. Purification of an endogenous proteininhibitor of the high affinity binding of gamma-aminobutyric acid to synapticmembranes of rat brain. *Proc Natl Acad Sci U S A.* 1978;75:4024–8.

Tonon MC, Vaudry H, Chuquet J, Guillebaud F, Fan J, Masmoudi-Kouki O, Vaudry D, Lanfray D, Morin F, Prevot V, Papadopoulos V, Troadec JD, Leprince J. Endozepines and their receptors: Structure, functions and pathophysiological significance. *Pharmacol Ther.* 2020;208:107386.

Truman JW. The evolution of insect metamorphosis. *Curr Biol.* 2019;29(23):R1252–68.

Wang JL, Wang JX, Zhao XF. Molecular cloning and expression profiles of the acyl-CoA-binding protein gene from the cotton bollworm *Helicoverpa armigera. Arch Insect Biochem Physiol.* 2008;68(2):79–88.

Wick JY. The history of benzodiazepines. *Consult Pharm.* 2013;28(9):538–48.

8 Biologically Active Phytochemicals Having Benzodiazepine-Like Actions

Domenico Montesano
University of Naples Federico II

Monica Gallo
University of Naples Federico II

CONTENTS

INTRODUCTION

To date, we can certainly consider anxiety as one of the main psychiatric disorders that cause suffering and damage in personal relationships. Furthermore, the treatment of this pathology represents a very high cost for the health system of each community as it negatively affects the quality of life with all the related consequences. Unfortunately, treatments for anxiety are not always effective, probably also due to the difficulty of intervening promptly with adequate diagnoses (Jiménez-Ferrera et al., 2017).

Anxiety is currently treated with two different approaches: pharmacotherapy and/or psychotherapy. In the first case, the well-known conventional synthetic anxiolytic drugs are used. Alternatively, or also in association with pharmacotherapy, psychotherapy is used, a nonpharmacological approach that relies on different techniques such as cognitive-behavior therapy, psychoanalysis, supportive counseling, brief psychotherapy, and group psychotherapy. Often, the use of anxiolytic drugs, especially those of the older generation, is accompanied by various adverse effects. Moreover, it is necessary to mention the difficulty of many people to undergo conventional psychological or pharmacological treatments as this is typically associated with psychiatric disorders, and still several individuals show intolerance and/or resilience to current drugs (Halberstein, 2005; Baldwin et al., 2010). We already have a lot of scientific evidence on the use of medicinal herbs for the treatment of mental disorders, and on this trial, new herbal medicines could represent an interesting option for anxiety and insomnia especially considering all the limitations of current conventional treatments. In line with this assumption, many researchers have dedicated themselves to finding novel safe and effective drugs from natural sources, and especially in recent years, there has been greater attention to natural supplements of plant origin as an alternative to the currently used synthetic drugs (Fedotova et al., 2017; Lakhan and Vieira, 2010). As shown by many experimental studies, biologically active compounds (BACs) or phytochemicals present in foods (such as alkaloids, terpenes, flavonoids, phenolic acids, lignans, cinnamates, and saponins or various plant extracts rich in these substances) have anxiolytic effects when tested in different animal models of anxiety. The ability of anxiolytic and hypno-sedative drugs such as benzodiazepines (BDZs) and barbiturates to regulate the level of inhibition of neurons is linked to their ability to bind allosteric sites on the GABAA receptor. Several BACs are able to interact with these allosteric sites (Figure 8.1) in the same way as synthetic drugs. These compounds, similar to BDZs, are capable of promoting changes in the conformation of the GABAA receptor by regulating the level of neuronal activity in states of pathological anxiety (Nuss, 2015).

All this takes on considerable importance, as in a context in which ~450 million people suffer from brain/mental or behavioral disorders, the possibility of having pharmaceutical products beyond conventional BDZs represents an urgent need.

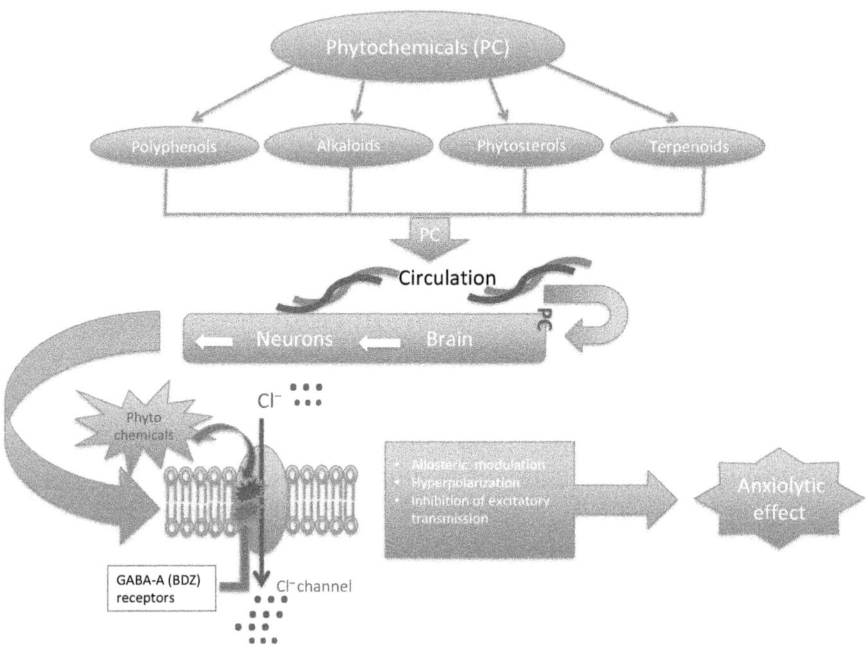

FIGURE 8.1 Phytochemicals interacting with the BDZ binding site on the GABAA receptor. Many BACs are able to bind to the BDZ binding site acting as positive allosteric modulators of the GABAA receptor, thus promoting the influx of Cl^- ions that causes membrane hyperpolarization and underlies their anxiolytic effect.

PLANTS AND PLANT PRODUCTS WITH ANXIOLYTIC-LIKE ACTIVITY

There are numerous works in the literature reporting the anxiolytic-like activity of plants or plant extracts, most of which performed on animal models. Many plants showing these properties belong to widely distributed and easily found botanical families such as *Lavandula angustifolia* L., *Achillea millefolium* L., *Humulus lupulus* L., or *Crocus sativus* L.

Lavandula angustifolia L. belongs to the *Lamiaceae* family, genus Lavandula, which includes evergreen shrubs widely distributed in the Mediterranean Basin, as well as throughout the Arabian Peninsula and in southern Asia (Mendez-Tovar et al., 2015). The cultivation of this species is carried out mainly for their essential oils, used in the field of perfumery, aromatherapy, and food. In fact, the knowledge of the phytochemistry of Lavandula is centered on the essential oils consisting mainly of mono- and sesquiterpenoid compounds (Harborne and Williams, 2002). The oil is generally obtained by steam distillation from the flowers of the plant, and to date, over 160 different substances have been identified. The anxiolytic properties of the essential oil have been attributed to different compounds but primarily to linalool and linalyl acetate (Setzer, 2009). In a study conducted on rodents by Kumar (2013), the anxiolytic effect of a preparation based on Lavender essential oil called "Silexan" (oil preparation in gelatin capsules containing 80 mg of lavender oil), administered intraperitoneally at doses of 3, 10, and 30 mg/kg for 7 days, was evaluated in rodents. The effects of this oil were compared with the effects of BDZs currently used for the treatment of anxiety, such as lorazepam and diazepam. The results of this study can be considered of great interest as dose-dependent anxiolytic effects were revealed after 1 h from the last administration of the natural drug using various experimental tests. The conclusion reached by the authors was that the observed anxiolytic effects were similar to those obtained using lorazepam. Preclinical data were also confirmed by clinical trials, whereby Silexan was shown to be safe and effective in patients with subthreshold anxiety or generalized anxiety disorder. Moreover, Silexan also exerted beneficial effects on symptoms commonly associated with anxiety, such as impaired sleep, somatic complaints, comorbid depression, or decreased quality of life. Interestingly, Silexan did not produce typical BDZ adverse effects, such as sedation or withdrawal symptoms, while mild gastrointestinal symptoms were the most relevant undesired effects at daily doses of 80 or 160 mg (Kasper et al., 2018).

In a research performed by Rahmati and collaborators (Rahmati et al., 2017), the effects of the hydroalcoholic extract of *L. officinalis* on scopolamine-induced memory impairment as well as on anxiety and depressive behaviors were evaluated. The experiment was conducted on rats treated with scopolamine alone or with lavender extracts at different concentrations 30 min before the intraperitoneal administration of the alkaloid. When scopolamine-induced memory impairment was assessed by Y-maze task, the groups of animals treated with lavender showed a better performance compared to the group treated with scopolamine alone. In addition to increasing the percentage of spontaneous alternation in the Y labyrinth in the groups pretreated with lavender, pretreatment with lavender at 200 and 400 mg/kg doses increased the time spent in the open arms, confirming an improvement in the disorders measured with these experimental assays.

Another botanical species of great interest for its anxiolytic properties is *Achillea millefolium* L. This plant, also known as yarrow, belongs to the Asteraceae family and is a perennial herb that has been used in folk medicine for hundreds of years in several countries. Its geographical origins can be considered the temperate regions of the Northern Hemisphere in Asia, Europe, and North America, and its spread is remarkable, going from the United Kingdom to China. Yarrow is currently considered one of the most used medicinal plants in the world: its active ingredients are effective for the treatment of wounds, to solve digestive problems, to fight respiratory infections and skin diseases, and show interesting properties as a mild sedative. Preclinical studies indicate that it may have anti-inflammatory, antiulcer, hepatoprotective, and anxiolytic effects (Applequist and Moerman, 2011). In a study performed on mice, yarrow was administered orally as a hydroalcoholic extract obtained from the aerial parts of the plant. The results of the experiment showed anxiolytic effects without changes in locomotor activity (Baretta et al., 2012).

The authors of this research noted that the anxiolytic effect of yarrow occurred after a chronic treatment for 25 days, thus inferring the absence of tolerance to this effect after short-term administration, such as for diazepam. Even more relevant were the binding data for [3H]-flunitrazepam according to which the interaction with the GABA/BDZ binding site does not mediate these effects, and this therefore differentiates *Achillea millefolium* L. from BDZs and other plant extracts with an anxiolytic effect in rodents (Duarte et al., 2008).

Humulus lupulus L. is another plant widely used in traditional medicine especially in European countries. This plant belongs to the *Cannabaceae* family, probably of Chinese origin, is a hardy and dioecious vine whose aerial part is herbaceous and annual (Astray et al., 2020). This species is commonly found in deciduous forests of Europe and western Asia, and generally grows in temperate climate zones (Muzykiewicz et al., 2019). In addition to the classic use as a bitter agent in the beer industry, hop cones have traditionally been applied for their multiple health properties and in particular for the control of spasms, anxiety, fever, inflammation, activation of the gastric function, and last but not least for the treatment of sleep disorders (Alonso-Esteban et al., 2019; Zanoli and Zavatti, 2008). In a study performed by Kyrou et al. (2017), the effects of a dry hop extract on self-reported levels of depression, anxiety, and stress in young adults were evaluated. Thirty-six adult participants completed the study intervention, and the results confirmed the interesting anxiolytic effects of the natural drug in accordance with what is reported in traditional medicine. Specifically, the authors noted a significant reduction in anxiety, depression, and stress in the hop-treated group compared to symptoms observed in the placebo group. In light of these experiments, the researchers performing this intervention study on young adults concluded that those who reported at least mild depression and stress symptoms, thanks to daily supplementation with a dry hop extract for 4 weeks, had a significant improvement of all these symptoms.

Much research has been devoted to saffron as a possible natural drug capable of relieving the symptoms of anxiety and depression. To date, saffron (*Crocus sativus* L.), a member of the *Iridaceae* family, can certainly be counted among the most expensive spices in the world. The precious part of the plant is represented by the dried stigma, which has been used as a spice for many years (Tarantilis and Polissiou, 1997). Saffron is a plant native to Iran, which still holds the primacy of world production today, although extensive crops can be found in many other temperate climate countries such as China, Spain, India, and Greece (Gohari et al., 2013). Several properties related to this precious spice are reported in the literature, such as the soothing capacity for stomach pain, or its ability to relieve the pain of renal colic, as a digestive, antispasmodic agent, as well as antidepressant (Akhondzadeh et al., 2005). The stigmas of saffron contain water-soluble carotenoids, crocins (Figure 8.2), monoterpenes such as picrotoxin and safranal in addition to flavonoid compounds, primarily quercetin and kaempferol (Tarantilis et al., 1995). Pitsikas et al. (2008) investigated the therapeutic potential of crocins in anxiety by means of an animal model, and in particular, they used the light/dark test on rats.

Furthermore, the treatment with crocin was evaluated in a locomotor activity test in order to distinguish between specific and nonspecific changes in animal activity. For this experiment, the animals were randomly divided into four experimental groups in addition to the control group supplemented with vehicle only. Three groups were treated with crocins (15, 30, 50 mg/kg) and one group with conventional reference drug, in this case diazepam (1.5 mg/kg). The results of this study show that when the crocins were administered at the highest dose (50 mg/kg), there was a significant increase in latency for the first transition to the dark side and in the time the rats spent in the illuminated chamber of the light/dark box. The diazepam-treated group showed nonstatistically different results. The treatment with crocins at lower doses (15 and 30 mg/kg) did not influence the behavior of the animals. In light of the results obtained from this research, the authors concluded that crocins were able to reduce the anxiety of animals exposed to the light/dark procedure without affecting the motor activity of the rats. Milajerdi et al. (2018) performed an 8-week double-blind

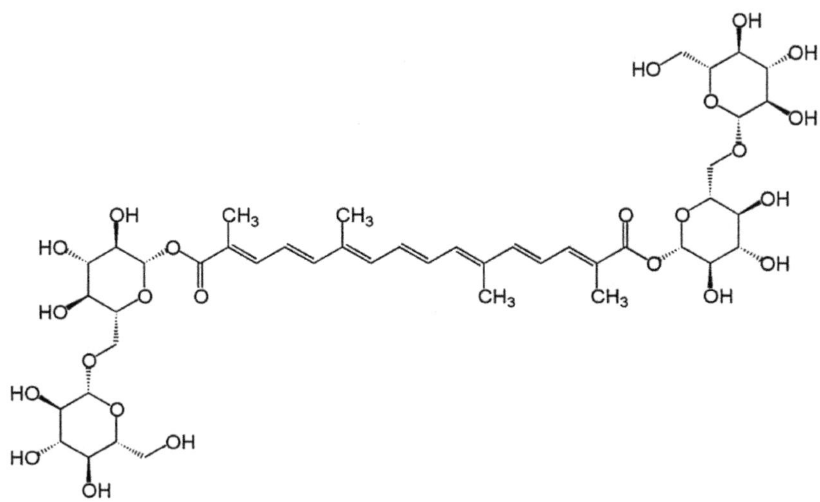

FIGURE 8.2 Chemical structure of carotenoid crocin-trans-crocetin di-(β-D-gentiobiosyl) ester.

randomized clinical trial in type-2 diabetic patients aged 40–65 who suffered from mild to moderate anxiety and depression. 54 patients were selected and randomly divided into two groups: one treated with placebo and the other with saffron extract twice a day for the duration of the experiment. Specifically, safrotin capsules were used as a pharmaceutical form for this study. Each saffron and placebo capsule contained 15 mg of hydroalcoholic extracts of saffron or placebo, respectively. In this particular experiment treating diabetic patients with sleep disturbance, anxiety, and depression, the consumption of hydroalcoholic extract of saffron stigma for 8 weeks led to a significant reduction in anxiety, depression plus anxiety scores, as well as a significant improvement in sleep disturbances after taking saffron for 8 weeks.

Kell et al. (2017) studied the effects of an innovative saffron extract (affron®) on mood disorders. The study was a parallel, double-blind, placebo-controlled design. Participants in this study were recruited by self-reported low mood albeit without a diagnosis of overt depression, and were randomly assigned to groups receiving saffron extract or placebo for 4 weeks. Each participant was recommended to take two tablets a day for the whole period of the experiment. The results of this research were very encouraging; in fact, there was a significant decrease in negative mood and symptoms related to stress and anxiety using a dose of 28 mg/day, while with the dose of 22 mg/day, no effect was recorded. The authors then concluded that affron® was able to improve mood, reducing anxiety and managing stress without side effects, so it can be considered an interesting natural alternative to conventional treatments.

Cucurbitaceae are among the botanical families whose species represent a rich reserve of BACs. *Citrullus colocynthis* (L.) Schrad. belongs to this family and is a medicinal plant known in traditional medicine for several beneficial properties, including the treatment of convulsions, or purgative, anti-inflammatory, antioxidant, antidiabetic, analgesic, lipid-lowering, and antiepileptic effects (Rahimi et al., 2012). The fruit of this plant contains various bioactive compounds, such as glycosides, flavonoids, alkaloids, carbohydrates, fatty acids, and essential oils (Jayaraman et al., 2009; Salama, 2012). Several studies report the ability of these BACs to interact with the GABA receptor. GABA is the main inhibitory neurotransmitter in mammals, responsible for regulating neuronal excitability and certainly involved in the etiopathogenesis of epilepsy, a syndrome that affects approximately 50 million people around the world (Moshi et al., 2005). Antiepileptic drugs act through different mechanisms to inhibit neuronal electrical signals and the excitatory activity of the epileptogenic foci. However, the therapy must be personalized and continued for long periods of time, often for a lifetime, and is not without side effects. Starting from these premises, the interest in exploring natural substances to counteract epileptic disease was aimed at identifying effective compounds with a limited toxicity. A group of Iranian researchers (Mehrzadi et al., 2016) evaluated the anticonvulsant activity and possible mechanisms of action of the pulp of the fruit of *C. colocynthis* in mice in which seizures had been induced with pentylenetetrazole, a noncompetitive antagonist at GABAA receptors. Results showed that pulp extract resulted in a statistically significant reduction in seizure duration and increased seizure latency period, in a dose-dependent manner at doses of 10, 25 and 50 mg/kg, but decreased at a dose of 100 mg/kg, probably due to its toxicity. Furthermore, the effects of naloxone and flumazenil on the anticonvulsant activity of *C. colocynthis* extract were evaluated. Naloxone, a synthetic drug used to block the effects of opioids, reduced the prolongation of seizure latency induced by *C. colocynthis* pulp extract; therefore, the anticonvulsant effect of *C. colocynthis* is likely due to an interaction with opioid receptors. Furthermore, the administration of flumazenil prior to *C. colocynthis* resulted in decreased seizure latency. The researchers observed that flumazenil was able to antagonize the effect of *C. colocynthis* on lag time seizures in the pentylenetetrazole model, so it is conceivable that the anticonvulsant effect was also due to the activation of BDZ receptors. The results of this research are compatible with the assumption that the pulp extract of *C. colocynthis* is able to delay the onset of convulsions induced by pentylenetetrazole and reduce their duration by interacting with the GABAergic system.

Verbena (scientific name: *Verbena officinalis* L.) is a perennial herb belonging to the *Verbenaceae* family, native to the humid temperate zones of Europe, Western Asia, and North Africa. This plant has been widely used in traditional folk medicine for the treatment of various diseases; in fact, it has been used as a sedative, relaxant, nervous tonic, gastroprotective, anti-inflammatory, analgesic, and antidepressant drug (Akanmu et al., 2002; Speroni et al., 2007). With regard to the aerial parts of this plant, effects against anxiety, insomnia, nervous irritability, and convulsions have also been reported (Lai et al., 2006; Braga et al., 2012). In a research by Rashidian et al. (2017), ethanolic extracts of the aerial parts of *V. officinalis* were used to evaluate the anticonvulsant activity and the probable involvement of the GABAergic and opioid systems. The trial was conducted on mice exposed to chemically and electrically induced seizures using pentylenetetrazole and maximum electroshock (MES), respectively. The results of this study confirmed the anticonvulsant effect of the extract; in fact, a dose-dependent protection with both of the above-mentioned models was detected. In particular, using MES, the extract proved effective in reducing the duration of tonic extension of the hind limbs. The analysis of the ethanolic extract of *V. officinalis* showed the presence of glycosides and alkaloids, which could be involved in the anticonvulsant activity of the plant.

Ziziphus jujuba Mill. var. spinosa (Bunge) Hu ex H. F. Chou, botanical name of Jujube, is a centenary plant belonging to the *Rhamnaceae* family, very resistant to diseases, which dates back to more than 4,000 years ago. *Ziziphus jujuba* is the fruit of the plant, while *Ziziphus spinosa* (ZS) are the seeds of this plant. In traditional Chinese medicine, the use of the seeds against insomnia has been recorded. Several effects of BACs present in ZS at the central nervous system level have been reported in the literature, such as neuronal protection (Park et al., 2004) and anxiolytic effects (Han et al., 2009). These effects seem related to the presence of different classes of phytochemicals in ZS, including alkaloids, flavonoids, triterpenes, and triterpene saponins (such as ziziberanalic acid, oleanolic acid, ursolic acid, jujubosides A and B) (Yang et al., 2013; Liu et al., 2007). Sedative and hypnotic activities of flavonoids, saponins, and polysaccharides are well documented in the literature (Shergis et al., 2017), and these properties are also demonstrated both *in vitro* and in animal models. Some studies have confirmed jujuboside A activity at the central level, detecting an inhibition of the excitatory state of the rat hippocampus through the glutamate-mediated excitatory signaling pathway (Shou et al., 2002; Zhang et al., 2003). When the effect of ZS is compared to that of diazepam, it appears to have slower action, longer duration with less impairment of learning and memory. This may be due to the effects of the aforementioned BACs on the GABAergic system and their ability to promote conformational changes of the GABAA receptor subunits (Ma et al., 2007, 2008). Accordingly, in cultured rat hippocampus neurons, it was shown that different doses of jujuboside A/jujubogenin promote changes in the gene expression of the GABAA receptor subunits (α1, α5, and ß2) (You et al., 2010). Furthermore, the ability of jujuboside A to induce the increased gene expression of GABAA receptor subunits (α1, α5, ß2) even at low doses was also highlighted, and it is likely that this may further enhance the sedative–hypnotic effects. These results are of particular interest, as they indicate that a jujuboside A drug strategy has potentially fewer side effects/undesirable effects in the long-term therapies than diazepam use. The alkaloid fraction of ZS exhibits interesting interactions with the GABAergic system. Sanjoinin A, a cyclopeptide alkaloid, behaves similarly to non-BDZ drugs (GABAA receptor agonists) in a dose-dependent manner and increases the influx of chloride into primary cultured cerebellar granular cells, causing changes in the GABAergic systems (Villanaueva and Villanaueva, 2017). ZS phytochemicals belonging to the flavonoid class have been shown to cure insomnia. The flavonoid 6-hydroxyflavone showed partial agonist activity thanks to its ability to bind to BDZ binding site on GABAA receptors (Ren et al., 2010). Spinosin is a glycosylated flavonoid with hypno-sedative properties, in fact when administered together with a 5-HT1A (5-hydroxytryptamine1A) receptor antagonist (4-(2-methoxy-phenyl)-1-[2-(*n*-2″-pyridinyl)-p-iodobenzamido]-ethyl-piperazine) increased total sleep time in rats treated with pentobarbital (Lee et al., 2016). To evaluate the anxiolytic effects of spinosin, Liu and co-workers used a behavioral test known as a light/dark box test as an animal model of anxiety. The mice were given either spinosin or diazepam orally prior to the behavioral test. The results obtained show that spinosin used at a dose of 5 mg/kg significantly increased the number of transitions between the compartments and the time spent in the light compartment. Therefore, it exerts anxiolytic-like effects and its mechanism of action appears to be modulated by GABAA and 5-HT1A receptors, antagonized, respectively, by flumazenil and WAY-100635 (Liu, 2015). Moreover, the effect of spinosin on the central nervous system is evidenced by its extensive distribution in regional brain tissue, particularly in the striatum and hippocampus (Zhang et al., 2015). For what has been described so far by reviewing the *in vitro* and *in vivo* studies related to ZS compounds, it seems evident that they have sedative and anxiolytic effects as GABA agonists.

The *Passiflora* genus includes more than 500 species, and several species have been widely used in folk medicine for their sedative and tranquilizing properties. *P. incarnata, P. edulis*, and *P. alata* are the most investigated species to which the anxiolytic/sedative properties are ascribed. These species appear as vines and have a great diffusion in tropical and subtropical regions especially in Central or South America, while they are not very present in North America, Southeast Asia, and Australia (Ulmer et al., 2004; Yuan et al., 2017). *P. incarnata* or "Passiflora" (*Flos passionis*) owes its name to the Spanish missionaries in South America who described its flower. "Passionis" refers to the passion of Christ, representing various elements of the crucifixion (Ratsch, 1998). As for *P. incarnata*, its use as a herbal medicine mainly acts as a mild sedative and anxiolytic. Currently, the dried aerial parts of the plant are those reported in the European and British Pharmacopoeias (Wohlmuth et al., 2010). The phytochemical analysis of the aerial parts of this species revealed the presence of flavonoids, maltol, cyanogenic glycosides, and indole alkaloids (Marchart et al., 2003; Krenn, 2001). As the chemical composition reported for *P. incarnata* is very rich and varied, so it is likely that the clinical effects ascribed to it are due to a synergistic action of the various classes of BACs. Among the compounds with an inhibitory action on monoamine oxidase (MAO), we must mention the indole alkaloids, represented as harman, harmin, harmaline, harmol, and harmalol (Sampath et al., 2011). The class of flavonoids is well represented with the presence, among others, of vitexin, isovitexin, orientin, isoorientine, apigenin, kaempferol, vicenin, lucenin, and saponarin (Miroddi et al., 2013; Marchart et al., 2003; Krenn, 2001). Several studies report that flavonoids can be metabolized by the intestinal microflora into their corresponding hydroxyphenylacetic acids; therefore, the polyphenolic structures present in nature could represent prodrugs that are activated after the intestinal passage and action of the microorganisms present there (Vissiennon et al., 2012). *P. incarnata* has been the subject of several preclinical and clinical studies in order to evaluate its anxiolytic and sedative effects. A Passionflower extract was used in a randomized double-blind clinical trial conducted on 36 patients with generalized anxiety disorder (GAD); its effect was compared to that obtained with oxazepam (Akhondzadeh et al., 2001). The results obtained from this study showed that both pharmacological treatments were effective in the treatment of GAD. Despite its rapid onset of action, oxazepam caused more relationship problems, also affecting the quality of work of people treated with this drug. Therefore, in light of these results, the authors concluded that Passionflower extract was an effective drug for the management of GAD with the advantage of a low incidence of impaired work performance compared to oxazepam. The sedative action of *P. incarnata* has also been tested in several experiments conducted on rodents. Grundman et al. (2008) investigated the anxiolytic properties of a commercial *P. incarnata* extract rich in flavonoids, using the elevated plus maze test in mice. The authors noted that the anxiolytic effects of passionflower extract at the dose of 375 mg/kg were antagonized by flumazenil but not by the specific 5HT1A-antagonist WAY-100 635. This is what currently happens for diazepam, for which it is assumed that the extract may share the same mechanism of action as BDZ, i.e., its interaction with the GABAergic system.

In recent and older literature, there is also scientific evidence in favor of the use of phytochemicals and products of plant origin for the treatment of anxiety, insomnia, depression, and cognitive decline. In all cases, the GABAergic system is always involved and the ability of phytochemicals to bind to GABA receptor sites is of fundamental importance, reproducing the mechanism of action of BDZs. Table 8.1 shows the most interesting botanical species with BDZ-like action.

The binding affinity for GABA receptor sites, their ability to modulate GABA neurotransmission, the interaction with monoamine MAO-A and MAO-B, as well as their ability to interfere with the serotonergic and noradrenergic systems, are some of the specific mechanisms by which the phytochemicals exert their actions on the CNS.

TABLE 8.1
The Most Common Plants with Proven Anxiolytic-Like Efficacy

Botanical Species	Medical Prescription	Mechanism of Action	Active Principles	References
Centella asiatica L.	Treatment of anxiety, cognitive impairment	GABA transaminase inhibition	Asiaticoside	Awad et al. (2007) and Wijeweera et al. (2006)
Cimicifuga racemosa L. Nutt	Treatment of anxiety and depression menopause	Action in the hypothalamus vasomotor center. Dopaminergic effect	Triterpenes and flavones	Dog et al. (2003) and McKenna et al. (2001)
Ginkgo biloba L.	Treatment of anxiety, depression, cognitive impairment	GABAergic effects; Modulation of cholinergic and monoamine pathways; Antioxidant	Isorhamnetin kaempferol, quercetin, ginkgoin and ginkgolin acids, ginkgolides A, B, C, and M and bilobalides	Di Renzo (2000) and Woelk et al. (2007)
Hypericum perforatum L.	Treatment of anxiety, insomnia, and depression	Selective inhibitor of monoamine oxidase (MAO)-A and MAO-B; Inhibition of serotonin (5-HT), norepinephrine (NE) and dopamine uptake; activity on GABAA receptor	Hypericin and hyperforin	Russo et al. (2013), Bilia et al. (2002), and Kiemow et al. (2011)
Matricaria recutita L.	Treatment of anxiety and insomnia	Interaction with GABA receptors; Modulation of monoamine neurotransmission	Apigenin, α-bisabolol	Zanoli et al. (2000) and Amsterdam et al. (2009)
Melissa officinalis L.	Treatment of anxiety	Inhibitor of rat brain GABA transaminase	Rosmarinic acid, triterpenoids, ursolic acid, and oleanolic acid	Taiwo et al. (2012), Kennedy et al. (2002) and Awad et al. (2009)

(*Continued*)

TABLE 8.1 (*Continued*)
The Most Common Plants with Proven Anxiolytic-Like Efficacy

Botanical Species	Medical Prescription	Mechanism of Action	Active Principles	References
Mimosa pudica L.	Treatment of depression and anxiety	Interaction with serotoninergic system	Norepinephrine, d-pinitol, β-sitosterol, mimosine	Guzmán-Gutiérrez (2014) and Zaware et al. (2014)
Piper methysticum G. Foster	Treatment of anxiety	Inhibition of MAO-B; blockade of NE uptake	Kawain, dihydrokavain, methysticin, dihydromethysticin, and yangonin	Sarris et al. (2013) and Seitz et al. (1997)
Rhodiola rosea L.	Treatment of anxiety and depression	MAO-A inhibition; normalization of 5-HT transmission	Salidroside, tyrosol Rosavin	Van Diermen et al. (2009), Chen et al. (2009), and Perfumi and Mattioli (2007)
Scutellaria lateriflora L.	Treatment of anxiety and insomnia	GABA-α binding	Scutelaterin A, baicalin	Awad et al. (2003)
Valeriana officinalis L.	Treatment of anxiety and insomnia	GABA modulation; 5-HT5A partial agonism	Valepotriate, valerenic acid	Benke et al. (2009), Bent et al. (2006), and Trauner et al. (2008)
Withania somnifera L.	Treatment of anxiety and insomnia	GABA-mimetic activity	Withanolide, withaferin	Battacharya and Muruganandam (2003), and Battacharya et al. (2000)

CONCLUDING REMARKS

Anxiety and mood disorders are currently the most widespread mental health problems worldwide. Various drugs are used for the treatment of these disorders, primarily BDZs, azapirone, tricyclic antidepressants, MAO inhibitors, selective serotonin reuptake inhibitors, norepinephrine reuptake inhibitors, serotonin and noradrenaline reuptake inhibitors, etc. However, there is not always a complete remission of symptoms. In addition, BDZs, i.e., the first-choice drugs in the treatment of anxiety and sleep disorders, have different side effects, including tolerance, and patients taking these drugs experience sedation which is incompatible with car driving and work that requires a high attention span. Considering therefore the importance of these disorders, their incidence on the general population and the various difficulties revealed by conventional drugs, the scientific world in the past 20 years has increasingly turned to the study of natural alternatives for their treatment, producing many data on the various phytochemicals and plants from which they derive. To date, phytomedicine seems increasingly involved in the treatment of anxiety, insomnia, and depression, also thanks to modern analytical tools for phytochemical investigation. On the other hand, the development of anxiolytic and antidepressant drugs of plant origin requires a multidisciplinary approach that includes not only an ethnopharmacological investigation, but also phytochemical, pharmacological, and biochemical studies. Actually, we have a lot of scientific evidence regarding many species of plants with central nervous system activity, so much that many of them are currently used in therapies to treat human diseases. From this point of view, similarly to many diseases treated with natural medicines, we could imagine the use of herbal medicines also for the treatment of anxiety and insomnia as an effective and safe option. In line with these considerations, many researchers are working to evaluate the anxiolytic activity of new natural plant compounds or plant extracts to provide health systems around the world with new weapons to combat these subtle and often disabling diseases. Currently, some drug strategies involve the use of different classes of phytochemicals as supplements to conventional anxiolytic therapies in order to improve efficacy and reduce adverse effects. Reading the numerous papers published on this topic, some of the most relevant presented in this chapter, the results appear to be encouraging, although further research is needed especially to elucidate the structure–activity relationships, metabolism, absorption, and neuro-psycho-pharmacological mechanisms of these phytochemicals and plant extracts. Although several studies have been conducted on humans, it must be borne in mind, however, that most of the research has been performed on animal models, so they require clinical evaluation in humans in order to validate the therapeutic effects of these new potential medications for the treatment of anxiety.

REFERENCES

Akanmu M.A., Honda K., Inoue S. 2002. Hypnotic effects of total aqueous extracts of *Vervain hastata* (Verbenaceae) in rats. *Psychiatry and Clinical Neurosciences* 56: 309–310.

Akhondzadeh S., Naghavi H., Vazirian M., Shayeganpour A., Rashidi H., Khani M. 2001. Passionflower in the treatment of generalized anxiety: a pilot double-blind randomized controlled trial with oxazepam. *Journal of Clinical Pharmacology and Therapeutics* 26: 363–367.

Akhondzadeh S., Tahmacebi-Pour N., Noorbala A.A., et al. 2005. *Crocus sativus* L. in the treatment of mild to moderate depression: a double-blind, randomized and placebo-controlled trial. *Phytotherapy Research* 19(2): 148–151.

Alonso-Esteban J.I., Pinela J., Barros L., et al. 2019. Phenolic composition and antioxidant, antimicrobial and cytotoxic properties of hop (*Humulus lupulus* L.) seeds. *Industrial Crops and Products* 134: 154–159.

Amsterdam J.D., Li Y., Soeller I., Rockwell K., Mao J.J., Shults J. 2009. A randomized, double-blind, placebo-controlled trial of oral *Matricaria recutita* (chamomile) extract therapy for generalized anxiety disorder. *Journal of Clinical Psychopharmacology* 29: 378–382.

Applequist W.L., Moerman D.E. 2011. Yarrow (*Achillea millefolium* L.): a neglected panacea? A review of ethnobotany, bioactivity, and biomedical research. *Economic Botany* 65(2): 209–225.

Astray G., Gullón P., Gullón B., Munekata P.E.S., Lorenzo J.M. 2020. *Humulus lupulus* L. as a natural source of functional biomolecules. *Applied Sciences* 10: 5074. doi:10.3390/app10155074.

Awad R., Arnason J.T., Trudeau V., et al. 2003. Phytochemical and biological analysis of skullcap (*Scutellaria lateriflora* L.): a medicinal plant with anxiolytic properties. *Phytomedicine* 10: 640–649.

Awad R., Levac D., Cybulska P., Merali Z., Trudeau V.L., Arnason J.T. 2007. Effects of traditionally used anxiolytic botanicals on enzymes of the gamma-aminobutyric acid (GABA) system. *Canadian Journal of Physiology and Pharmacology* 85: 933–942.

Awad R., Muhammad A., Durst T., Trudeau V.L., Arnason J.T. 2009. Bioassay-guided fractionation of lemon balm (*Melissa officinalis* L.) using an in vitro measure of GABA transaminase activity. *Phytotherapy Research* 23: 1075–1081.

Baldwin D.S., Ajel K.I., Garner M. 2010. Pharmacological treatment of generalized anxiety disorder. *Current Topics in Behavioral Neurosciences* 2: 453–467.

Baretta P., Felizardo R.A., Bimbato V.F., et al. 2012. Anxiolytic-like effects of acute and chronic treatment with *Achillea millefolium* L. extract. *Journal of Ethnopharmacology* 140: 46–54.

Benke D., Barberis A., Kopp S., et al. 2009. GABA(A) receptors as in vivo substrate for the anxiolytic action of valerenic acid, a major constituent of valerian root extracts. *Neuropharmacology* 56: 174–181.

Bent S., Padula A., Moore D., Patterson M., Mehling W. 2006. Valerian for sleep: a systematic review and meta-analysis *American Journal Medicine* 119: 1005–1012.

Bhattacharya S.K., Bhattacharya A., Sairam K., Ghosal S. 2000. Anxiolytic-antidepressant activity of *Withania somnifera* glycowithanolides: an experimental study. *Phytomedicine* 7: 463–469.

Bhattacharya S.K., Muruganandam A.V. 2003. Adaptogenic activity of *Withania somnifera*: an experimental study using a rat model of chronic stress. *Pharmacology Biochemistry & Behavior* 75: 547–555.

Bilia A.R., Gallori S., Vincieri F.F. 2002. St. John's wort and depression efficacy, safety and tolerability-an update. *Life Sciences* 70: 3077–3096.

Braga V.F., Mendes G.C., Oliveira R.T., et al. 2012. Micropropagation, anti-nociceptive and antioxidant activities of extracts of Verbena litoralis Kunth (Verbenaceae). *Anais da Academia Brasileira de Ciências* 84: 139–148.

Chen Q.G., Zeng Y.S., Qu Z.Q., Tang J.Y., Qin Y.J., Chung P., Wong R., Hagg U. 2009. The effects of Rhodiola rosea extract on 5-HT level, cell proliferation and quantity of neurons at cerebral hippocampus of depressive rats. *Phytomedicine* 16: 830–838.

Di Renzo G. 2000. Ginkgo biloba and the central nervous system. *Fitoterapia* 71 (Suppl 1): S43–S47.

Dog L.T., Powell L.K., Weisman M.S. 2003. Critical evaluation of the safety of *Cimicifuga racemosa* in menopause symptom relief. *Menopause Journal* 10: 299–313.

Duarte F.S., Marder M., Hoeller A.A. 2008. Anticonvulsant and anxiolytic-like effects of compounds isolated from *Polygala sabulosa* (Polygalaceae) in rodents: in vitro and in vivo interactions with benzodiazepine binding sites. *Psychopharmacology (Berl)* 197: 351–360.

Fedotova J., Kubatkad P., Büsselberg D., et al. 2017. Therapeutical strategies for anxiety and anxiety-like disorders using plant-derived natural compounds and plant extracts. *Biomedicine & Pharmacotherapy* 95: 437–446.

Gohari A.R., Saeidnia S., Mahmoodabadi M.K. 2013. An overview on saffron, phytochemicals, and medicinal properties. *Pharmacognosy Review* 7(13):61–66.

Grundmann O., Wang J., McGregor G.P., Butterweck V. 2008. Anxiolytic activity of a phytochemically characterized Passiflora incarnata extract is mediated via the GABAergic system. *Planta Medica* 74: 1769–1773.

Guzmán-Gutiérrez S.L., Reyes-Chilpa R., Bonilla-Jaime H. 2014. Medicinal plants for the treatment of "nervios", anxiety, and depression in Mexican Traditional Medicine. *Revista Brasileira de Farmacognosia* 24: 591–608.

Halberstein R.A., 2005. Medicinal plants: historical and cross-cultural usage patterns. *Annals of Epidemiology* 15(9): 686–699.

Han H., Ma Y., Eun J.S., et al. 2009. Anxiolytic-like effects of sanjoinine A isolated from Zizyphi spinosi semen: possible involvement of GABAergic transmission. *Pharmacology Biochemistry and Behavior* 92: 206–213.

Harborne J.B., Williams C.A. 2002. Phytochemistry of the genus Lavandula. In: *Lavender.* Lis-Balchin, M. (Ed.) London and New York: Taylor & Francis, pp. 86–99.

Jayaraman R., Shivakumar A., Anitha T., Joshi V., Palei N. 2009. Antidiabetic effect of petroleum ether extract of *Citrullus colocynthis* fruits against streptozotocin-induced hyperglycemic rats. *Romanian Journal of Biology* 4: 127–134.

Jiménez-Ferrera E., Santillán-Urquiza M.A., Alegría-Herrera E., et al. 2017. Anxiolytic effect of fatty acids and terpenes fraction from Aloysia triphylla: serotoninergic, GABAergic and glutamatergic implications. *Biomedicine & Pharmacotherapy* 96: 320–327.

Kasper S, Müller WE, Volz HP, Möller HJ, Koch E, Dienel A. 2018. Silexan in anxiety disorders: clinical data and pharmacological background. *World J Biol Psychiatry* 19(6): 412–420.

Kell G., Rao A., Beccaria G., Clayton P., Inarejos-García A.M., Prodanov M. 2017. Affron® a novel saffron extract (*Crocus sativus* L.) improves mood in healthy adults over 4 weeks in a double-blind, parallel, randomized, placebo-controlled clinical trial. *Complementary Therapies in Medicine* 33: 58–64.

Kennedy D.O., Scholey A.B., Tildesley N.T., Perry E.K., Wesnes K.A. 2002. Modulation of mood and cognitive performance following acute administration of Melissa officinalis (lemon balm). *Pharmacology Biochemistry & Behavior* 72: 953–964.

Klemow K.M., Bartlow A., Crawford J., Kocher N., Shah J., Ritsick M. 2011. Medical attributes of St. John's Wort (Hypericum perforatum), in: *Herbal Medicine: Biomolecular and Clinical Aspects.* Benzie I.F.F., Sissi W.G., (Eds.) 2nd edn, Boca Raton (FL): CRC Press/Taylor & Francis, pp. 1–27.

Krenn L. 2001. Passion Flower (*Passiflora incarnata* L.) – a reliable herbal sedative. *Wiener Medizinische Wochenschrift* 152 (15–16): 404–406.

Kumar V. 2013. Characterization of anxiolytic and neuropharmacological activities of Silexan. *Wiener Medizinische Wochenschrift* 163: 89–94.

Kyrou I., Christou A., Panagiotakos D., et al. 2017. Effects of a hops (*Humulus lupulus* L.) dry extract supplement on self-reported depression, anxiety and stress levels in apparently healthy young adults: a randomized, placebo-controlled, double-blind, crossover pilot study. *Hormones* 16(2):171–180.

Lai S.W., Yu M.S., Yuen W.H., Chang R.C.C. 2006. Novel neuroprotective effects of the aqueous extracts from *Verbena officinalis* Linn. *Neuropharmacology* 50: 641–650.

Lakhan S.E., Vieira K.F. 2010. Nutritional and herbal supplements for anxiety and anxiety-related disorders: systematic review. *Nutrition Journal* 9: 42.

Lee Y., Jeon S., Lee H., Jung I. 2016. Spinosin, a C-glycoside flavonoid, enhances cognitive performance and adult hippocampal neurogenesis in mice. *Pharmacology Biochemistry & Behavior* 145: 9–16.

Liu J. 2015. GABA and 5-HT systems are implicated in the anxiolytic-like effect of spinosin in mice. *Pharmacology Biochemistry & Behavior* 128: 41–49.

Liu J., Chen B., Yao S. 2007. Simultaneous analysis and identification of main bioactive constituents in extract of *Zizyphus jujuba* var. sapinosa (Zizyphi spinosi semen) by high-performance liquid chromatography–photodiode array detection–electrospray mass spectrometry. *Talanta* 71: 668–675.

Ma Y., Han H., Eun J.S., Kim H.C., Hong J.T., Oh, K.W. 2007. Sanjoinine A isolated from Zizyphi spinosi semen augments pentobarbital-induced sleeping behaviors through the modification of GABA-ergic systems. *Biological and Pharmaceutical Bulletin* 30: 1748–1753.

Ma Y., Han H., Nam S.Y., et al. 2008. Cyclopeptide alkaloid fraction from Zizyphi spinosi semen enhances pentobarbital-induced sleeping behaviors. *Journal of Ethnopharmacology* 117: 318–324.

Marchart E., Krenn L., Kopp B. 2003. Quantification of the flavonoid glycosides in *Passiflora incarnata* by capillary electrophoresis. *Planta Medica* 69: 452–466.

McKenna D.J., Jones K., Humphrey S., Hughes K. 2001. Black cohosh: efficacy, safety, and use in clinical and preclinical applications. *Alternative Therapies in Health and Medicine* 7: 93–100.

Mehrzadi S., Shojaii A., Pur S.A., Motevalian M. 2016. Anticonvulsant activity of hydroalcoholic extract of *Citrullus colocynthis* fruit: involvement of benzodiazepine and opioid receptors. *Journal of Evidence-Based Complementary & Alternative Medicine* 21(4): NP31-NP35.

Mendez-Tovar I., Herrero B., Pérez-Magariño S., Pereira J. A., Asensio-S.-Manzanera M. C. 2015. By-product of Lavandula latifolia essential oil distillation as source of antioxidants. *Journal of Food and Drug Analysis* 23: 225–233.

Milajerdi A., Jazayeri S., Shirzadi E., Hashemzadeh N., Azizgol A., Djazayery A. 2018. The effects of alcoholic extract of saffron (*Crocus satious* L.) on mild to moderate comorbid depression-anxiety, sleep quality, and life satisfaction in type 2 diabetes mellitus: a double-blind, randomized and placebo-controlled clinical trial. *Complementary Therapies in Medicine* 41: 196–202.

Miroddi M., Calapai G., Navarra M., Minciullo P.L., Gangemi S. 2013. *Passiflora incarnata* L.: ethnopharmacology, clinical application, safety and evaluation of clinical trials. *Journal of Ethnopharmacology* 150: 791–804.

Moshi M.J., Kagashe G.A., Mbwambo Z.H. 2005. Plants used to treat epilepsy by Tanzanian traditional healers. *Journal of Ethnopharmacol*ogy 97:327–336.

Muzykiewicz A., Nowak A., Zielonka-Brzezicka J., Florkowska K., Duchnik W., Klimowicz A. 2019. Comparison of antioxidant activity of extracts of hop leaves harvested in different years. *Herba Polonica* 65: 1–9.

Nuss P. 2015. Anxiety disorders and GABA neurotransmission: a disturbance of modulation. *Neuropsychiatric Disease and Treatment* 11: 165–175.

Park J., Lee H., Koh S., Ban J., Seong Y. 2004. Protection of NMDA-induced neuronal cell damage by methanol extract of Zizyphi spinosi semen in cultured rat cerebellar granule cells. *Journal of Ethnopharmacology* 95: 39–45.

Perfumi M., Mattioli L. 2007. Adaptogenic and central nervous system effects of single doses of 3% rosavin and 1% salidroside *Rhodiola rosea* L. extract in mice. *Phytotherapy Research* 21: 37–43.

Pitsikas N., Boultadakis A., Georgiadou G., Tarantilis P.A., Sakellaridis N. 2008. Effects of the active constituents of *Crocus sativus* L., crocins, in an animal model of anxiety. *Phytomedicine* 15: 1135–1139.

Rahimi R., Amin G., Ardekani MR. 2012. A review on *Citrullus colocynthis* Schrad.: from traditional Iranian medicine to modern phytotherapy. *Journal of Alternative and Complementary Medicine* 18(6): 551–554. doi: 10.1089/acm.2011.0297.

Rahmati B., Kiasalari Z., Roghani M., Khalili M., Ansari F. 2017. Antidepressant and anxiolytic activity of *Lavandula officinalis* aerial parts hydroalcoholic extract in scopolamine-treated rats. *Pharmaceutical Biology* 55 (1): 958–965.

Rashidian A., Kazemi F., Mehrzadi S., Dehpour A.R., Mehr S.E., Rezayat S.M. 2017. Anticonvulsant effects of aerial parts of *Verbena officinalis* extract in mice: involvement of benzodiazepine and opioid receptors. *Journal of Evidence-Based Complementary & Alternative Medicine* 22(4): 632–636.

Ratsch C. 1998. *The Encyclopedia of Psychoactive Plants: Ethnopharmacology and Its Applications.* Rochester, VT: Park Street Press.

Ren L., Wang F., Xu Z., Chan W. 2010. GABAA receptor subtype selectivity underlying anxiolytic effect of 6-hydroxyflavone. *Biochemical Pharmacology* 79: 1337–1344.

Russo E., Scicchitano F., Whalley B.J., et al. 2013. Hypericum perforatum: pharmacokinetic, mechanism of action, tolerability, and clinical drug-drug interactions. *Phytotherapy Research* 28: 643–655.

Salama H.M. 2012. Alkaloids and flavonoids from the air dried aerial parts of *Citrullus colocynthis. Journal of Medicinal Plant Research* 6: 5150–5155.

Sampath C., Holbik M., Krenn L., Butterweck V. 2011. Anxiolytic effects of fractions obtained from *Passiflora incarnata* L. in the elevated plus maze in mice. *Phytotherapy Research* 25: 789–795.

Sarris J., Stough C., Bousman C.A., et al. 2013. Kava in the treatment of generalized anxiety disorder: a double-blind, randomized, placebo-controlled study. *Journal of Clinical Psychopharmacology* 33: 643–648.

Seitz U., Schulle A., Gleitz J. 1997. [3H]-monoamine uptake inhibition properties of kava pyrones. *Planta Medica* 63: 548–549.

Setzer W.N. 2009. Essential oils and anxiolytic aromatherapy. *Natural Product Communications* 4: 1305–1316

Shergis J. L., Ni X., Sarris J., et al. 2017. *Ziziphus spinosa* seeds for insomnia: a review of chemistry and psychopharmacology. *Phytomedicine* 34: 38–43.

Shou C., Feng Z., Wang J., Zheng X. 2002. The inhibitory effects of jujuboside A on rat hippocampus in vivo and in vitro. *Planta Medica* 68: 799–803.

Speroni E., Cervellati R., Costa S., et al. 2007. Effects of differential extraction of *Verbena officinalis* on rat models of inflammation, cicatrization and gastric damage. *Planta Medica* 73: 227–235.

Taiwo A.E., Leite F.B., Lucena G.M., et al. 2012. Anxiolytic and antidepressant-like effects of *Melissa officinalis* (lemon balm) extract in rats: influence of administration and gender. *Indian Journal of Pharmacology* 44: 189–192.

Tarantilis P.A., Polissiou M.G. 1997. Isolation and identification of the aroma components from saffron (*Crocus sativus*). *Journal of Agricultural and Food Chemistry* 45(2): 459–462.

Tarantilis P.A., Tsoupras G., Polissiou M. 1995. Determination of saffron (*Crocus sativus* L.) components in crude plant extract using high-performance liquid chromatography UV/visible photodiode-array detection-mass spectrometry. *Journal of Chromatography* 699: 107–118.

Trauner G., Khom S., Baburin I., Benedek B., Hering S., Kopp B. 2008. Modulation of GABAA receptors by valerian extracts is related to the content of valerenic acid. *Planta Medica* 74: 19–24.

Ulmer T., MacDougal J.M., Ulmer B. 2004. *Passiflora: Passionflowers of the World.* Portland, OR: Timber Press.

Van Diermen D., Marston A., Bravo J., Reist M., Carrupt P.A., Hostettmann K. 2009. Monoamine oxidase inhibition by *Rhodiola rosea* L. roots. *Journal of Ethnopharmacology* 122: 397–401.

Villanaueva J., Villanaueva L. 2017. Experimental and clinical pharmacology of *Ziziphus jujuba* mills. *Phytotherapy Research* 31: 347–365.

Vissiennon C., Nieber K., Kelber O., Butterweck V. 2012. Route of administration determines the anxiolytic activity of the flavonols kaempferol, quercetin and myricetin–are they prodrugs? *Journal of Nutritional Biochemistry* 23: 733–740.

Wijeweera P., Arnason J.T., Koszycki D., Merali Z. 2006. Evaluation of anxiolytic properties of Gotukola — (*Centella asiatica*) extracts and asiaticoside in rat behavioral models. *Phytomedicine* 13(9–10): 668–676.

Woelk H., Arnoldt K., Kieser M., Hoerr R. 2007. Ginkgo biloba special extract EGb 761 in generalized anxiety disorder and adjustment disorder with anxious mood: a randomized, doubleblind, placebo-controlled trial. *Journal of Psychiatric Research* 41: 472–480.

Wohlmuth H., Penman K.G., Pearson T., Lehmann R.P. 2010. Pharmacognosy and chemotypes of passionflower (*Passiflora incarnata* L.). *Biological and Pharmaceutical Bulletin* 33: 1015–1018.

Yang B., Yang H., Chen F., Hua Y., Jiang Y. 2013. Phytochemical analyses of *Ziziphus jujuba* mill. var. spinosa seed by ultrahigh performance liquid chromatography-tandem mass spectrometry and gas chromatography-mass spectrometry. *Analyst* 138: 6881–6888.

You Z., Xia Q., Liang F.-R., et al. 2010. Effects on the Expression of GABAA receptor subunits by jujuboside a treatment in rat hippocampal neurons. *Journal of Ethnopharmacology* 128: 419–423.

Yuan T.Z., Kao C.L., Li W.J., Li H.T., Chen C.Y. 2017. Chemical constituents of leaves of *Passiflora edulis. Chemistry of Natural Compounds* 53: 1165–1166. doi:10.1007/s10600-017-2227-5

Zanoli P., Avallone R., Baraldi M. 2000. Behavioral characterisation of the flavonoids apigenin and chrysin. *Fitoterapia* 71(Suppl 1): S117–S123.

Zanoli P., Zavatti M. 2008. Pharmacognostic and pharmacological profile of *Humulus lupulus* L. *Journal of Ethnopharmacology* 116: 383–396.

Zaware B.B., Chaudhari S.R., Shinde M.T. 2014. An overview of *Mimosa pudica* Linn.: chemistry and pharmacological profile. *Research Journal of Pharmaceutical, Biological and Chemical Sciences* 5: 754–761.

Zhang M., Ning G., Shou C., Lu Y., Hong D., Zheng X. 2003. Inhibitory effect of jujuboside A on glutamate-mediated excitatory signal pathway in hippocampus. *Planta Medica* 69: 692–695.

Zhang Y., Zhang T., Wang F., Xie J. 2015. Brain tissue distribution of spinosin in rats determined by a new high-performance liquid chromatography-electrospray ionization–mass/mass spectrometry. *Journal of Chromatography Science* 53: 97–103.

9 Benzodiazepines, Flavonoids, and GABA

Tina Hinton and Graham A. R. Johnston
The University of Sydney

CONTENTS

INTRODUCTION

Benzodiazepines (both endogenous and synthetic) allosterically modulate the action of GABA, the major inhibitory neurotransmitter in the brain, on $GABA_A$ receptors (Hinton and Johnston 2018). Such allosteric modulators are much sought-after for their potential subtype selectivity as a result of the greater structural diversity of allosteric sites as distinct from the orthosteric GABA binding site (Changeux 2013; Christopoulos 2002). While benzodiazepines are the most widely studied modulators, many non-benzodiazepines have potent actions as modulators (Solomon et al. 2019). Of particular interest are flavonoids that were described early on as "a new family of benzodiazepine receptor ligands" (Medina et al. 1997).

Flavonoids occur naturally in many plant products. Mammals do not make flavonoids; rather, flavonoids are consumed regularly in the diet where they are known to cross the blood–brain barrier to influence brain function (Spencer 2008; Kozlowska and Szostak-Wegierek 2014; Johnston 2015; Williamson et al. 2018).

Dietary intake of flavonoids in US adults has been estimated to be some 340 mg/day, 190 mg/day being flavan-3-ols (Bai et al. 2014). Tea, wine, beer, citrus fruits, and apples were the most important sources of total flavonoid intake in the US adults.

While natural and synthetic flavonoids act as antioxidants and have structurally specific actions on many different systems, there is particular interest in their potent actions on $GABA_A$ receptors. Benzodiazepines and flavonoids appear to interact with overlapping binding sites on certain $GABA_A$ receptors.

GABA RECEPTORS

There are two types of GABA receptors in the brain. Ionotropic GABA receptors are ligand-gated ion channels that, when activated, normally allow chloride ions to flow into the neuron, thus inhibiting neuronal firing. Metabotropic GABA receptors are G protein-coupled receptors that influence neuronal function via second messengers.

Ionotropic GABA receptors are pentameric structures made up of five protein subunits that confer specific physiological and pharmacological properties. $GABA_A$ receptors are heteromeric being made up of two or three different protein subunit subclasses, the most common combination being two α-, two β-, and one γ-subunits. $GABA_A$ receptors are modulated by a wide variety of substances. Positive modulators increase the action of GABA in activating the receptors, while negative modulators reduce such action. Silent modulators block the action of both positive and negative modulators without directly influencing the action of GABA itself. $GABA_C$ receptors are homomeric being made up of five ρ-subunits. $GABA_C$ receptors are insensitive to the convulsant alkaloid bicuculline, whereas $GABA_A$ receptors are antagonized by bicuculline (Johnston 2013). They are more sensitive to GABA than most $GABA_A$ receptors and are not modulated by the wide variety of substances that modulate $GABA_A$ receptors, including benzodiazepines. $GABA_C$ receptors have been described as relatively simple transmitter-gated ion channels (Johnston 1996).

BENZODIAZEPINES AND $GABA_A$ RECEPTORS

There is a vast literature on the effects of benzodiazepines on $GABA_A$ receptors. Detailed studies have been made on the benzodiazepine binding sites on the various subtypes of $GABA_A$ receptors (Sigel and Ernst 2018). Benzodiazepines are often described in terms of agonists and antagonists at $GABA_A$ receptors, but in fact they are more aptly described as modulators. Allosteric receptor modulation is now recognized in terms such as positive allosteric modulator (PAM), negative allosteric modulator (NAM), and silent allosteric modulator (SAM) that are used to describe different types of modulator. SAMs are also called neutralizing allosteric modulators. There are also allosteric agonists, and second-order modulators have been described that modulate the action of first-order modulators such as benzodiazepines (Hinton and Johnston 2018).

"Benzodiazepine agonists," such as diazepam, are PAMs enhancing the action of GABA on these receptors, while "benzodiazepine antagonists," such as flumazenil, are SAMs that block the modulatory action of benzodiazepine PAMs (Figure 9.1). SAMs also block NAMs such as methyl β-carboline-3-carboxylate (Vignes et al. 2006).

Benzodiazepines have no direct action on $GABA_A$ receptors. Instead, the PAMs enhance the action of GABA by increasing the frequency of ion channel openings. Benzodiazepines bind at the interface of the α- and γ-subunits. The α-subunit requires a histidine residue as in α1-, α2-, α3-, and α5-subunits. $GABA_A$ receptors containing α1-subunits are associated with stronger hypnotic and sedative effects of

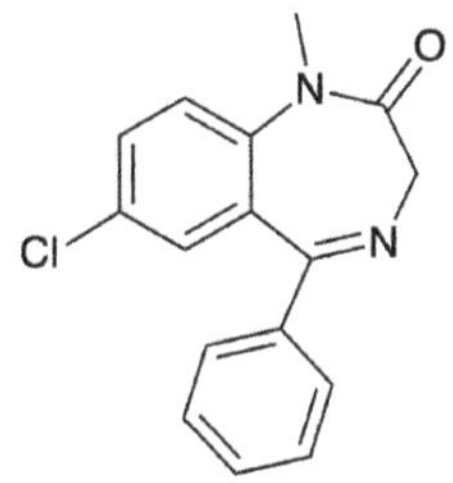

Diazepam, a synthetic benzodazepine
Positive allosteric modulator of $GABA_A$ receptors acting on high and low affinity sites
Anxiolytic, sedative, muscle relaxant

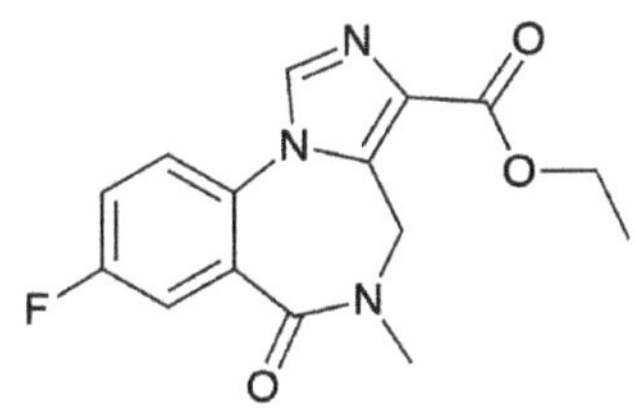

Flumazenil, a synthetic benzodiazepine
Silent allosteric modulator of the high affinity action of diazepam on $GABA_A$ receptors

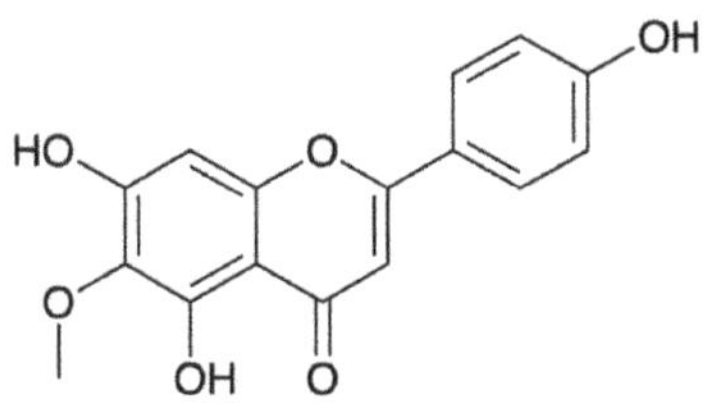

Hispidulin, a flavone found in sage
Positive allosteric modulator of $GABA_A$ receptors
Partially flumazenil sensitive
Anticonvulsant

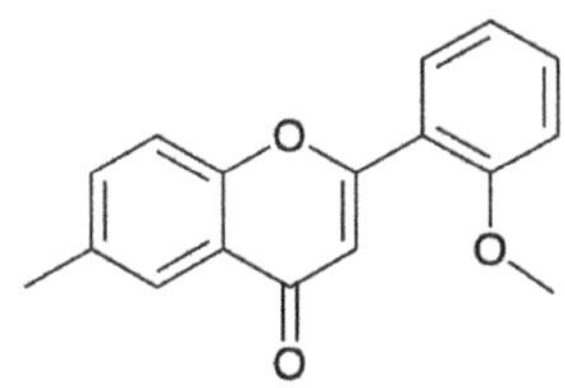

2'-Methoxy-6-methylflavone, a synthetic flavone
Flumazenil-insensitive positive allosteric modulator of most $GABA_A$ receptors.
Direct activator of α2β2/3γ2L $GABA_A$ receptors
Anxiolytic, sedative
Neuroprotection after cerebral ischaemia

FIGURE 9.1 Structures of representative benzodiazepines and flavonoids.

benzodiazepines, while those containing α2- and α3-subunits are stronger anxiolytics (Fritschy 2015; Crestani and Rudolph 2015).

Endogenous benzodiazepines ("endozepines"), such as the 86-amino acid polypeptide diazepam binding inhibitor (DBI), have been shown to have both PAM and NAM actions on $GABA_A$ receptors (Tonon et al. 2020).

It has been shown that benzodiazepines act on $GABA_A$ receptors via two distinct and separable mechanisms (Walters et al. 2000). Studies using recombinant α1β2γ2 $GABA_A$ receptors with low concentrations of GABA showed a biphasic potentiation by diazepam with distinct components in the nanomolar (high-affinity) and micromolar (low-affinity) range. Mutations in the second transmembrane domains of the subunits abolished the low-affinity component. The low-affinity component was independent of the γ2-subunit and was insensitive to the benzodiazepine SAM flumazenil. The authors suggested that the low-affinity component might be associated with the anesthetic actions of non-benzodiazepines such as midazolam and lorazepam.

Flumazenil-insensitive benzodiazepine binding sites in $GABA_A$ receptors contribute to diazepam-induced immobility in zebrafish (Cao et al. 2019). The authors described these as nonclassical benzodiazepine binding sites. Flumazenil-insensitive benzodiazepine effects have been found in a variety of $GABA_A$ receptor subtypes and may be important for designing of subtype- or binding site-specific agents (Lian et al. 2020).

FLAVONOIDS AND $GABA_A$ RECEPTORS

Flavonoids are polyphenolic plant products that occur in many foods and beverages. The pioneering studies on naturally occurring and synthetic flavonoids carried out by the research groups of Marder, Medina, Paladini in Argentina during the 1990s drew attention to flavonoids initially as "a new family of benzodiazepine receptor ligand" (Medina et al. 1997; Paladini et al. 1999). They have now been studied extensively for their effects on $GABA_A$ receptors (Hinton et al. 2017). They can act as PAM, NAM, and SAM. Some actions are benzodiazepine-like and also flumazenil-sensitive, whereas others are flumazenil-insensitive. They can directly activate $GABA_A$ receptors acting as allosteric agonists. They can act to modulate the action of other modulators. This ability to influence function via their actions on $GABA_A$ receptors permits a range of effects of flavonoids, including relief of anxiety, anticonvulsant, analgesic, and sedative actions. They are being explored for the treatment of neuropathic pain (Uddin et al. 2020).

Many flavonoids elicit biphasic responses on GABA action on $GABA_A$ receptors, enhancing GABA actions at low concentrations and inhibiting at high concentrations. Additionally, some flavonoids act as agonists particularly at high concentrations and directly gate the receptor in the absence of GABA (Chua et al. 2015). Clearly then, flavonoids interact with at least two and possibly more specific active sites on $GABA_A$ receptors (Hinton et al. 2017). Some of these sites appear to be independent of the classical flumazenil-sensitive benzodiazepine sites.

Flumazenil-Sensitive Actions of Flavonoids

Both naturally occurring and synthetic flavonoids have potent flumazenil-sensitive PAM actions on $GABA_A$ receptors. For example, 5,7,2'-trihydroxy-6,8-dimethoxyflavone, a natural anxiolytic found in the medicinal herb *Scutellaria baicalensis Georgi*, is as potent as diazepam in its flumazenil-sensitive action on $GABA_A$ receptors (Huen et al. 2003). Another naturally occurring flavonoid 6-hydroxyflavone shows flumazenil-sensitive subtype selectivity on $GABA_A$ receptors consistent with its anxiolytic actions (Ren et al. 2010). The synthetic flavonoid 6,2'-dihydroxyflavone is a flumazenil-sensitive NAM, showing subtype specificity consistent with its anxiogenic and cognitive-enhancing actions (Wang et al. 2007).

Extensive quantitative structure–activity studies have been carried out on flavonoids and their interactions with the flumazenil-sensitive benzodiazepine sites on $GABA_A$ receptors with a view to develop a comprehensive pharmacophore model to design new agents acting at these sites (Deeb et al. 2014; Dekermendjian et al. 1999).

Some flavonoids, however, have mixed actions with respect to flumazenil sensitivity. For example, the flavone hispidulin (Figure 9.1) is a potent PAM of $GABA_A$ receptor that crosses the blood–brain barrier acting as an anticonvulsant (Kavvadias et al. 2004). Its action on recombinant $GABA_A$ receptors is only partially blocked by flumazenil, indicating an additional action on flumazenil-insensitive sites (Johnston and Beart 2004).

Such actions are not restricted to flavonoids. A wide range of plant-derived natural products act on $GABA_A$ receptors in flumazenil-sensitive and flumazenil-insensitive ways. These include alkaloids, terpenes, phenolic acids, lignans, and saponins, which act as anxiolytics (Fedotova et al. 2017). Indeed, a very wide chemical range of $GABA_A$ allosteric modulators do not act on the classical flumazenil-sensitive benzodiazepine sites (Solomon et al. 2019).

Flumazenil-Insensitive Actions of Flavonoids

Many flavonoids modulate the action of GABA on $GABA_A$ receptors in a flumazenil-insensitive manner. For example, 6-methylflavone is a flumazenil-insensitive PAM of the action of GABA at $GABA_A$ receptors acting on α1β2γ2 recombinant receptors that have high-affinity sites for benzodiazepines and α1β2 receptors that do not have such sites (Hall et al. 2004). The anxiolytic and sedative actions of the flavonoid baicalein involve flumazenil-insensitive actions on $GABA_A$ receptors (de Carvalho et al. 2011). Prenylated flavonoids, such as xanthohumol, mediate their sedative action as PAMs of $GABA_A$ receptors in a flumazenil-insensitive manner (Benkherouf et al. 2019). Amentoflavone, a NAM of the action of GABA on $GABA_A$ receptors, is also insensitive to flumazenil (Hanrahan et al. 2003).

Of particular interest is 2'-methoxy-6-methylflavone (Figure 9.1), a synthetic flavone that acts as a flumazenil-insensitive PAM of most α1-containing $GABA_A$ receptors (Karim et al. 2012). It also directly activates α2β2/3γ2L $GABA_A$ receptors in the absence of GABA. Mutation studies showed that position 265 in the β1/2-subunit was key to whether 2'-methoxy-6-methylflavone was an activator or a potentiator. In mice, it showed anxiolytic and sedative effects without myorelaxation. Its pharmacological profile is similar to that of the general anesthetic etomidate dependent on different subunit stoichiometries of $GABA_A$ receptor subtypes (Chua et al. 2015). Its action on δ-containing $GABA_A$ receptors afforded neuroprotection following cerebral ischemia and shows a promise in the treatment of stroke (Clarkson et al. 2019).

Thus, rather than being regarded as a "new family of benzodiazepine receptor ligands," flavonoids should be viewed as significant allosteric modulators and activators of $GABA_A$ receptors in their own right.

Flavonoid Effects on Benzodiazepine Allosteric Modulation of $GABA_A$ Receptors

The dietary flavonoids apigenin and (-)-epigallocatechin gallate (EGCG, Figure 9.2) enhance the PAM of diazepam on recombinant $GABA_A$ receptors; i.e., they modulate the modulator (Campbell et al. 2004). This effect is observed at flavonoid concentrations that do not influence GABA action in the absence of diazepam. It is specific to diazepam modulation, not being observed with positive modulation by allopregnanolone or pentobarbitone.

Apigenin and EGCG have been described as second-order allosteric modulators of the first-order modulation of $GABA_A$ receptors by benzodiazepines (Hinton et al. 2017). The observed second-order modulation may result from alterations in the coupling of the flumazenil-sensitive benzodiazepine allosteric sites with the orthosteric GABA sites on $GABA_A$ receptors. In order for this to be a mechanism for the inhibition of locomotor activity by apigenin, there would have to be primary modulation by endogenous benzodiazepines (Tonon et al. 2020).

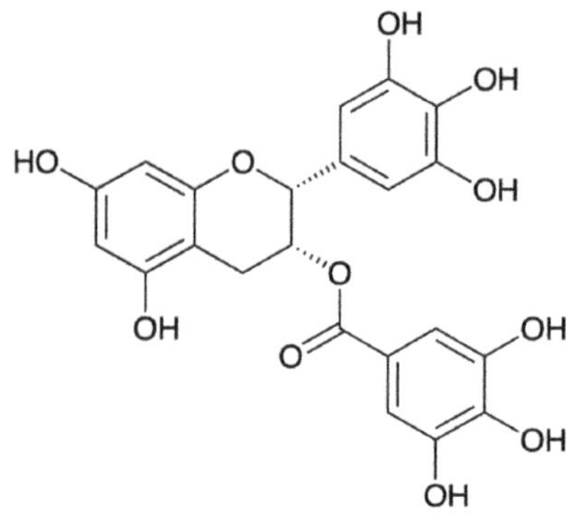

Epigallocatechin gallate (EGCG), a flavan-3-ol found in green tea
Second order positive allosteric modulator of primary
allosteric modulator diazepam on $GABA_A$ receptors.
Also counteracts negative allosteric modulation
by methyl β-carboline-3-carboxylate
Anxiolytic, sedative hypnotic

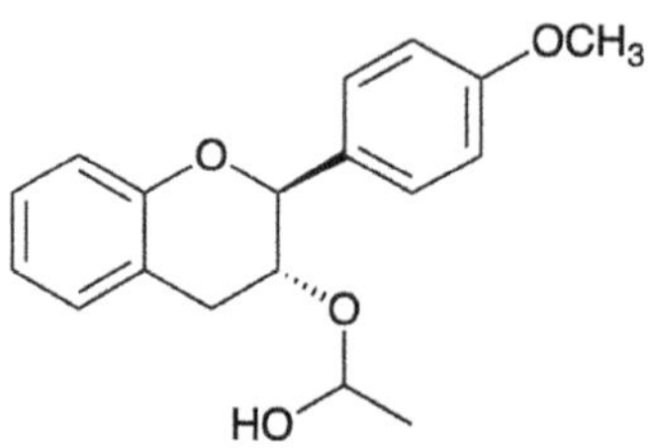

Fa131, a synthetic flavan-3-ol
Positive allosteric modulator
of $GABA_A$ receptors
Flumazenil-insensitive
Selective for α2 over α1, α3, α5
Non-sedating anxiolytic

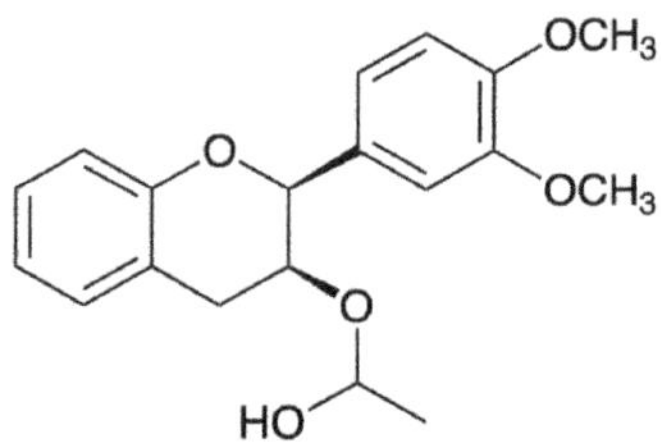

Fa173, a synthetic flavan-3-ol
Silent allosteric modulator of Fa131,
etomidate, loreclezole and
low but not high affinity diazepam.
Inactive against propofol, thiopental
and allopregnanolone

FIGURE 9.2 Structures of flavan-3-ols that influence the modulation of $GABA_A$ receptors.

EGCG (Figure 9.2) is the major polyphenol in green tea that exhibits dose-dependent stress-reducing, anxiolytic, sedative, and hypnotic properties. It exhibits its second-order modulation at low concentrations (0.1 mM), while at higher concentrations (>1 mM) it inhibits the action of GABA (Campbell et al. 2004). At low concentrations, EGCG is able to reverse the NAM action of ethyl β-carboline (Vignes et al. 2006). This indicates that EGCG may act as a second-order modulator with respect to the first-order modulation by both positive and negative modulators that act via benzodiazepine-binding sites on $GABA_A$ receptors (Hinton et al. 2017).

Second-order modulation (or metamodulation) has also been noted in other systems (Mesce 2002; Ribeiro and Sebastiao 2010) and may represent an obscure novel mechanism of drug action deserving further investigation, with the potential to lead to decreased therapeutic doses. It is not easy to investigate as it involves the dose-dependent interactions between three ligands, requiring the study of a number of dose combinations, and thus may be difficult to observe. Synergistic interactions have been described between other flavonoids on GABA receptor-related behaviors and at glycine receptors between strychnine and flavonoids (Hinton et al. 2017). Complex tertiary interactions between flavonoids and other substances may be a subtle feature of Cys-loop ligand-gated ion channels.

Flavan-3-ol Esters

EGCG is a relatively complex flavan-3-ol ester. In structure–activity studies on EGCG, a small library of simpler chemicals has been synthesized (Mewett et al. 2009). Two such analogues of EGCG proved to be of particular interest: Fa131 and Fa173 (Fernandez et al. 2012).

Fa131 (trans-(2S,3R)-3-acetoxy-4′-methoxyflavan, Figure 9.2) is a non-sedating anxiolytic and a selective flumazenil-insensitive PAM of α2-containing $GABA_A$ receptors (Fernandez et al. 2008). Fa131 is the first PAM to distinguish between α2- and α3-containing $GABA_A$ receptors, highlighting the potential of targeting flumazenil-insensitive allosteric sites in the search for new anxio-selective drugs.

Fa173 (cis-(2S,3S)-3-acetoxy-3′,4′-dimethoxyflavan, Figure 9.2), a diastereo-isomeric flavan-3-ol ester related to Fa131 with an additional methoxy in the 3′ position, was shown to block the modulatory actions of Fa131 (Fernandez et al. 2012). Additionally, Fa173 also blocked the enhancement of the GABA response by the anesthetic etomidate, the sedative anticonvulsant loreclezole, and selectively blocked the low-affinity effect of diazepam (100 μM) at α1β2γ2L and α1β2 $GABA_A$ receptors, but not the high-affinity effect of diazepam (100 nM). Fa173 was the first SAM of the low-affinity flumazenil-insensitive benzodiazepine binding site.

Fa173 was found not to inhibit the PAM of GABA by the anesthetic propofol, barbiturate thiopental, or neuroactive steroid allopregnanolone. This suggests that Fa131, etomidate, loreclezole, and high (non-flumazenil-sensitive) doses of diazepam all exert their PAM effects via a common or overlapping binding site that can be blocked by the silent modulator Fa173 (Fernandez et al. 2012).

Perhaps, the Fa173-sensitive sites on $GABA_A$ receptors would be more appropriately described as flavonoid sites as they appear to be activated by many naturally occurring and synthetic flavonoids (Fernandez et al. 2012).

CONCLUSION

Benzodiazepines and flavonoids act on specific sites on the $GABA_A$ receptor complex leading to anxiolysis and sedation. These actions may be divided into two classes by flumazenil, a SAM of $GABA_A$ receptors. The flumazenil-sensitive sites are modulated by nM concentrations of benzodiazepines, whereas the flumazenil-insensitive sites require μM concentrations. Flavonoids may act on either of the classes with relatively high affinity. The action of flavonoids on the flumazenil-insensitive sites can be blocked by the SAM Fa173, a synthetic flavonoid. It might be appropriate to think of the two classes of modulatory sites on $GABA_A$ receptors as benzodiazepine and flavonoid sites, respectively, which are flumazenil-sensitive and Fa173-sensitive.

Flavonoids represent a structurally distinct class of allosteric modulators of $GABA_A$ receptors from the benzodiazepines. Flavonoids thus have a distinct pharmacological profile from endogenous or synthetic benzodiazepines. As many flavonoids occur in the diet, they may be found in the brain where they are able to modulate the function of the major inhibitory neurotransmitter.

REFERENCES

Bai, W., C. Wang, and C. Ren. 2014. "Intakes of total and individual flavonoids by US adults." *International Journal of Food Sciences and Nutrition* 65 (1):9–20. doi: 10.3109/09637486.2013.832170.

Benkherouf, A. Y., S. L. Soini, M. Stompor, and M. Uusi-Oukari. 2019. "Positive allosteric modulation of native and recombinant $GABA_A$ receptors by hops prenylflavonoids." *European Journal of Pharmacology* 852:34–41. doi: 10.1016/j.ejphar.2019.02.034.

Campbell, E. L., M. Chebib, and G. A. R. Johnston. 2004. "The dietary flavonoids apigenin and (-)-epigallocatechin gallate enhance the positive modulation by diazepam of the activation by GABA of recombinant $GABA_A$ receptors." *Biochemical Pharmacology* 68 (8):1631–1638. doi: 10.1016/j.bcp.2004.07.022.

Cao, Y. Q., H. Yan, G. Yu, and R. B. Su. 2019. "Flumazenil-insensitive benzodiazepine binding sites in $GABA_A$ receptors contribute to benzodiazepine-induced immobility in zebrafish larvae." *Life Sciences* 239:17033–17033. doi: 10.1016/j.lfs.2019.117033.

Changeux, J. P. 2013. "The concept of allosteric modulation: an overview." *Drug Discov Today Technol* 10 (2):e223–e228. doi: 10.1016/j.ddtec.2012.07.007.

Christopoulos, A. 2002. "Allosteric binding sites on cell-surface receptors: novel targets for drug discovery." *Nature Reviews Drug Discovery* 1 (3):198–210. doi: 10.1038/nrd746.

Chua, H. C., N. L. Absalom, J. R. Hanrahan, R. Viswas, and M. Chebib. 2015. "The direct actions of GABA, 2'-methoxy-6-methylflavone and general anaesthetics at beta3gamma2l $GABA_A$ receptors: evidence for receptors with different subunit stoichiometries." *PLoS ONE [Electronic Resource]* 10 (10):e0141359.

Clarkson, A. N., L. Boothman-Burrell, Z. Dosa, R. Y. Nagaraja, L. Jin, K. Parker, P. S. van Nieuwenhuijzen, S. Neumann, E. K. Gowing, N. Gavande, P. K. Ahring, M. M. Holm, J. R. Hanrahan, J. A. Nicolazzo, K. Jensen, and M. Chebib. 2019. "The flavonoid, 2'-methoxy-6-methylflavone, affords neuroprotection following focal cerebral ischaemia." *Journal of Cerebral Blood Flow & Metabolism* 39 (7):1266–1282. doi: 10.1177/0271678X18755628.

Crestani, F., and U. Rudolph. 2015. "Behavioral functions of $GABA_A$ receptor subtypes--the Zurich experience." *Advances in Pharmacology* 72:37–51. doi: 10.1016/bs.apha.2014.10.001.

de Carvalho, R. S. M., F. S. Duarte, and T. C. M. de Lima. 2011. "Involvement of GABAergic non-benzodiazepine sites in the anxiolytic-like and sedative effects of the flavonoid baicalein in mice." *Behavioural Brain Research* 221 (1):75–82. doi: 10.1016/j.bbr.2011.02.038.

Deeb, O., B. Shaik, and V. K. Agrawal. 2014. "Exploring QSARs of the interaction of flavonoids with $GABA_A$ receptor using MLR, ANN and SVM techniques." *Journal of Enzyme Inhibition and Medicinal Chemistry* 29 (5):670–676. doi: 10.3109/14756366.2013.839557.

Dekermendjian, K., P. Kahnberg, M. R. Witt, O. Sterner, M. Nielsen, and T. Liljefors. 1999. "Structure-activity relationships and molecular modeling analysis of flavonoids binding to the benzodiazepine site of the rat brain $GABA_A$ receptor complex." *Journal of Medicinal Chemistry* 42 (21):4343–4350. doi: 10.1021/jm991010h.

Fedotova, J., P. Kubatka, D. Busselberg, A. G. Shleikin, M. Caprnda, J. Dragasek, L. Rodrigo, M. Pohanka, I. Gasparova, V. Nosal, R. Opatrilova, T. Qaradakhi, A. Zulli, and P. Kruzliak. 2017. "Therapeutical strategies for anxiety and anxiety-like disorders using plant-derived natural compounds and plant extracts." *Biomedicine & Pharmacotherapy* 95:437–446. doi: 10.1016/j.biopha.2017.08.107.

Fernandez, S. P., N. Karim, K. N. Mewett, M. Chebib, G. A. R. Johnston, and J. R. Hanrahan. 2012. "Flavan-3-ol esters: new agents for exploring modulatory sites on $GABA_A$ receptors." *British Journal of Pharmacology* 165 (4):965–977. doi: 10.1111/j.1476-5381.2011.01615.x.

Fernandez, S. P., K. N. Mewett, J. R. Hanrahan, M. Chebib, and G. A. R. Johnston. 2008. "Flavan-3-ol derivatives are positive modulators of $GABA_A$ receptors with higher efficacy for the alpha(2) subtype and anxiolytic action in mice." *Neuropharmacology* 55 (5):900–907. doi: 10.1016/j.neuropharm.2008.06.069.

Fritschy, J. M. 2015. "Significance of $GABA_A$ receptor heterogeneity: clues from developing neurons." *Advances in Pharmacology* 73:13–39.

Hall, B. J., M. Chebib, J. R. Hanrahan, and G. A. R. Johnston. 2004. "Flumazenil-independent positive modulation of gamma-aminobutyric acid action by 6-methylflavone at human recombinant alpha(1)beta(2)gamma(2L)- and alpha(1)beta(2) GABA(A) receptors." *European Journal of Pharmacology* 491 (1):1–8. doi: 10.1016/j.ejphar.2004.03.014.

Hanrahan, J. R., M. Chebib, N. M. Davucheron, B. J. Hall, and G. A. R. Johnston. 2003. "Semisynthetic preparation of amentoflavone: a negative modulator at $GABA_A$ receptors." *Bioorganic & Medicinal Chemistry Letters* 13: 2281–2284.

Hinton, T., J. Hanrahan, and G. A. R. Johnston. 2017. "Flavonoid actions on receptors for the inhibitory neurotransmitter GABA." *Flavonoids - from Biosynthesis to Human Health*, edited by J. Justino, InTech, Croatia, 335–349.

Hinton, T., and G. A. R. Johnston. 2018. "GABA, the major inhibitory neurotransmitter in the brain." *Reference Module in Biomedical Sciences*:1–15. doi: https://doi.org/10.1016/B978-0-12-801238-3.96594-2.

Huen, M. S. Y., K. M. Hui, J. W. C. Leung, E. Sigel, R. Baur, J. T. F. Wong, and H. Xue. 2003. "Naturally occurring 2 '-hydroxyl-substituted flavonoids as high-affinity benzodiazepine site ligands." *Biochemical Pharmacology* 66 (12):2397–2407. doi: 10.1016/j.bcp.2003.08.016.

Johnston, G. A. R. 1996. "$GABA_C$ receptors: relatively simple transmitter-gated ion channels?" *Trends in Pharmacological Sciences* 17 (9):319–323.

Johnston, G. A. R. 2013. "Advantages of an antagonist: bicuculline and other GABA antagonists." *British Journal of Pharmacology* 169 (2):328–336. doi: 10.1111/bph.12127.

Johnston, G. A. R. 2015. "Flavonoid nutraceuticals and ionotropic receptors for the inhibitory neurotransmitter GABA." *Neurochemistry International* 89:120–125. doi: 10.1016/j.neuint.2015.07.013.

Johnston, G. A. R., and P. M. Beart. 2004. "Flavonoids: some of the wisdom of sage?" *British Journal of Pharmacology* 142 (5):809–810. doi: 10.1038/sj.bjp.0705827.

Karim, N., J. Curmi, N. Gavande, G. A. R. Johnston, J. R. Hanrahan, M. L. Tierney, and M. Chebib. 2012. "2'-Methoxy-6-methylflavone: a novel anxiolytic and sedative with subtype selective activating and modulating actions at $GABA_A$ receptors." *British Journal of Pharmacology* 165 (4):880–896. doi: 10.1111/j.1476-5381.2011.01604.x.

Kavvadias, D., P. Sand, K. A. Youdim, M. Z. Qaiser, C. Rice-Evans, R. Baur, E. Sigel, W. D. Rausch, P. Riederer, and P. Schreier. 2004. "The flavone hispidulin, a benzodiazepine receptor ligand with positive allosteric properties, traverses the blood-brain barrier and exhibits anticonvulsive effects." *British Journal of Pharmacology* 142 (5):811–820. doi: 10.1038/sj.bjp.0705828.

Kozlowska, A., and D. Szostak-Wegierek. 2014. "Flavonoids--food sources and health benefits." *Roczniki Panstwowego Zakladu Higieny* 65 (2):79–85.

Lian, J. J., Y. Q. Cao, Y. L. Li, G. Yu, and R. B. Su. 2020. "Flumazenil-insensitive benzodiazepine effects in recombinant alpha beta and neuronal $GABA_A$ receptors." *Brain Sciences* 10 (3):13. doi: 10.3390/brainsci10030150.

Medina, J. H., H. Viola, C. Wolfman, M. Marder, C. Wasowski, D. Calvo, and A. C. Paladini. 1997. "Overview - flavonoids: a new family of benzodiazepine receptor ligands." *Neurochemical Research* 22 (4):419–425. doi: 10.1023/a:1027303609517.

Mesce, K. A. 2002. "Metamodulation of the biogenic amines: second-order modulation by steroid hormones and amine cocktails." *Brain Behavior and Evolution* 60 (6):339–349. doi: 10.1159/000067793.

Mewett, K. N., S. P. Fernandez, A. K. Pasricha, A. Pong, S. O. Devenish, D. E. Hibbs, M. Chebib, G. A. R. Johnston, and J. R. Hanrahan. 2009. "Synthesis and biological evaluation of flavan-3-ol derivatives as positive modulators of $GABA_A$ receptors." *Bioorganic & Medicinal Chemistry* 17 (20):7156–7173. doi: 10.1016/j.bmc.2009.08.062.

Paladini, A. C., M. Marder, H. Viola, C. Wolfman, C. Wasowski, and J. H. Medina. 1999. "Flavonoids and the central nervous system: from forgotten factors to potent anxiolytic compounds." *Journal of Pharmacy and Pharmacology* 51 (5):519–526. doi: 10.1211/0022357991772790.

Ren, L. H., F. Wang, Z. W. Xu, W. M. Chan, C. Y. Zhao, and H. Xue. 2010. "$GABA_A$ receptor subtype selectivity underlying anxiolytic effect of 6-hydroxyflavone." *Biochemical Pharmacology* 79 (9):1337–1344. doi: 10.1016/j.bcp.2009.12.024.

Ribeiro, J. A., and A. M. Sebastiao. 2010. "Modulation and metamodulation of synapses by adenosine." *Acta Physiologica* 199 (2):161–169. doi: 10.1111/j.1748-1716.2010.02115.x.

Sigel, E., and M. Ernst. 2018. "The benzodiazepine binding sites of $GABA_A$ receptors." *Trends in Pharmacological Sciences* 39 (7):659–671. doi: 10.1016/j.tips.2018.03.006.

Solomon, V. R., V. J. Tallapragada, M. Chebib, G. A. R. Johnston, and J. R. Hanrahan. 2019. "GABA allosteric modulators: an overview of recent developments in non-benzodiazepine modulators." *European Journal of Medicinal Chemistry* 171:434–461. doi: 10.1016/j.ejmech.2019.03.043.

Spencer, J. P. 2008. "Flavonoids: modulators of brain function?" *British Journal of Nutrition* 99 (E Suppl 1):ES60–ES77.

Tonon, M. C., H. Vaudry, J. Chuquet, F. Guillebaud, J. J. Fan, O. Masmoudi-Kouki, D. Vaudry, D. Lanfray, F. Morin, V. Prevot, V. Papadopoulos, J. D. Troadec, and J. Leprince. 2020. "Endozepines and their receptors: structure, functions and pathophysiological significance." *Pharmacology & Therapeutics* 208: 107386.

Uddin, M. S., A. Al Abdullah, M. A. Rahman, M. T. Kabir, S. Alkahtani, I. S. Alanazi, A. Perveen, G. M. Ashraf, M. N. Bin-Jumah, and M. M. Abdel-Daim. 2020. "Exploring the promise of flavonoids to combat neuropathic pain: from molecular mechanisms to therapeutic implications." *Frontiers in Neuroscience* 14: 478–478. doi: 10.3389/fnins.2020.00478.

Vignes, M., T. Maurice, F. Lante, M. Nedjar, K. Thethi, J. Guiramand, and M. Recasens. 2006. "Anxiolytic properties of green tea polyphenol (-)-epigallocatechin gallate (EGCG)." *Brain Research* 1110: 102–115. doi: 10.1016/j.brainres.2006.06.062.

Walters, R. J., S. H. Hadley, K. D. W. Morris, and J. Amin. 2000. "Benzodiazepines act on $GABA_A$ receptors via two distinct and separable mechanisms." *Nature Neuroscience* 3 (12):1274–1281. doi: 10.1038/81800.

Wang, F., Z. W. Xu, C. T. Yuen, C. Y. Chow, Y. L. Lui, S. Y. Tsang, and H. Xue. 2007. "6,2 '-dihydroxyflavone, a subtype-selective partial inverse agonist of $GABA_A$ receptor benzodiazepine site." *Neuropharmacology* 53 (4):574–582. doi: 10.1016/j.neuropharm.2007.06.018.

Williamson, G., C. D. Kay, and A. Crozier. 2018. "The bioavailability, transport, and bioactivity of dietary flavonoids: a review from a historical perspective." *Comprehensive Reviews in Food Science and Food Safety* 17 (5):1054–1112. doi: 10.1111/1541-4337.12351.

10 The Microbiome and Benzodiazepines: A Connection?

Jo Ann LeQuang
NEMA Research, Inc.

CONTENTS

INTRODUCTION

Virtually every mucosal and cutaneous surface of the human body is colonized by microbes, primarily but not exclusively bacteria, such that the average human adult hosts about 100 trillion microbiota cells, outnumbering human cells by at least a factor of ten.[1] Yet to consider the microbiome simply a large amount of microbial organisms occupying a specific space is to discount their importance, which is emerging as a vital new field of medical research. The host and the microbiome together form an ecological system based on mutually beneficial interactions and energetics.[2–4] Among other things, the human microbiome modulates responses to xenobiotics, exogenous drugs, and endogenous substances with the role fluctuating between that of watchman, communicating with distal organs, and gatekeeper, protecting the body from pathogens and toxins.[2,5–7] The extent and scope of the influence of the microbiome on human biology has yet to be fully elucidated.

The body contains multiple microbiomes, for instance in the mouth, on the skin, in the gut, in the vagina, and so on, but it is unclear if and to what extent the microbiota of these different regions contribute to the function of the others.[8] Some of the microbes preserved in the human microbiome might be considered "heirloom taxa," or bacteria that have been preserved over multiple human generations.[9] Others have described the microbiome as a metagenome, because it possesses most, if not all, of the genes related to drug metabolism. Unlike other genes expressed in humans, the genes of the microbiome are dynamic and respond to changes in diet, infections, health conditions, drug use, chemical agents, pollutants and toxins, and lifestyle factors.[10] It is known, for example, that the microbiome changes markedly during pregnancy, with aging, and with the use of certain drugs such as antibiotics.[11]

Current research is investigating the potential contribution of the microbiome to a variety of conditions, including obesity,[12] autism spectrum disorder,[13] chronic obstructive pulmonary disease,[14] cancer,[15] and attention deficit disorder.[16] Perhaps the best known microbiome of the human body resides in the gastrointestinal (GI) tract and makes up what is often called "the immune system." While the gut microbiome can affect how drugs are metabolized, it is actually far more complicated. For one thing, there is a bidirectional relationship between drug and microbiome, since the microbiome affects how the drug is metabolized but the drug affects the microbiome as well. The role of the microbiome in benzodiazepine metabolism—and the adverse effects, dependence, and withdrawal symptoms associated with benzodiazepine use—remains largely speculative. Certainly, our understanding of the microbiome is in its infancy. Considering that benzodiazepines are among the most frequently consumed drugs in the world and that there is a vast interpatient variability in their use, the microbiome may be an important avenue of investigation.

BENZODIAZEPINE METABOLISM

Long before, microbiomes were topics of scientific investigation; the pharmacological concepts of drug metabolism were spelled out in the familiar acronym of ADME: absorption, distribution, metabolism, and excretion. Originally, these four items were considered a linear progression, a step-by-step path of how the body processed exogenous drugs. Today, it is known that the microbiome plays a role in these steps. As with many other pharmacological agents, oral exogenous benzodiazepines undergo chemical modifications in the body that can result in the production of metabolites, which may possess functional and possibly toxicological properties distinct from those of the parent drug. Since benzodiazepines represent a large drug class with numerous individual agents each with their own distinct pharmacological properties, blanket statements about benzodiazepine metabolism cannot be made.[17] For example, diazepam, clonazepam, and midazolam are metabolized by the liver, produce active metabolites, and are characterized by a relatively long duration of action. In contrast, lorazepam, oxazepam, and temazepam produce no active metabolites and are metabolized via glucuronidation.

What can be said about the entire class of benzodiazepines is that they act as positive allosteric modulators of the gamma-aminobutyric acid (GABA) receptors. GABA receptors are inhibitory neurotransmitters that exert a calming, anxiolytic effect. Benzodiazepines are able to increase the flow of chloride ions through cell membranes, which, in turn, results in neuronal hyperpolarization and less frequent firing, meaning that anxiety-producing regions of the brain are less active. The patient may experience a sense of calm or a freedom from racing runaway thoughts. Since benzodiazepines are able to amplify and increase the inhibitory effects of GABA, they confer anxiolysis on the patient. For patients with overactive GABA receptors, neural circuits become overactive and can lead to somatic stress, anxiety, and worry. This effect may be situational, such as a patient dealing with sudden unexpected bad news, or it may arise with no particular context. The inhibition of GABA receptors with benzodiazepines produces a relaxing effect that can help a patient manage a short-term period of stress but may be less beneficial for longer-term use.[9]

THE GUT MICROBIOME

The human GI tract is home to trillions of microbes, most which are obligate anaerobes from the genera *Bacteroides*, *Clostridium*, *Lactobacillus*, *Escherichia*, and *Bifidobacteria*. In addition, there may be as many as 2,000 other species, including yeasts, viruses, and other microorganisms residing in the human gut, but less is known about the role of these organisms. The gut microbiota have both direct and indirect effects on how drugs are metabolized.[18,19] The gut microbiome has many functions: processing nutrients from food, aiding digestion, developing immune cells, modulating immune responses, resisting pathogens, controlling host energy, overseeing metabolic processes, and regulating brain development.[1] The gut microbiome produces specific factors for a variety of neuronal systems, such as neurotransmitters and neuromodulators (dopamine, noradrenaline, acetylcholine, GABA, and others),[20,21] forging what has been called the "gut–brain axis." Indeed, dysbiosis of the GI tract has been associated with central nervous system symptomology.[1]

The gut–brain axis relies on four main bidirectional pathways (Figure 10.1).[20,22–24] The first is the immune system pathway, which allows the microbiome in the GI tract to regulate the production and modulate the balance of cytokines, cytokinetic reaction factors, and prostaglandin E2.[25] This immune pathway encompasses the brain and, indeed, appears to have an effect on cerebral function. The neuroendocrine pathway is next, which helps to regulate the hypothalamic–pituitary–adrenal axis together with the central nervous system by adjusting the levels of cortisol, tryptophan, and serotonin.[26] The third path involves the vagus nerve, which appears to act as a conduit between gut and brain.[27] The role of the vagus pathway was described in a murine study. The bacteria *Lactobacillus* and *Bifidobacteria* secrete GABA,[28] and mice fed *Lactobacillus rhamnosus* had less anxiety and greater GABA receptor expression than control animals, similar to the effect of exogenous benzodiazepines.[29,30] However, this behavior did not occur in mice whose vagus nerve had been severed, suggesting that the vagus nerve forms a crucial pathway between the brain and the gut.

THE GUT MICROBIOME AND BENZODIAZEPINE METABOLISM

Among the metabolically relevant actions that can be attributed to the gut microbiome are acetylation and diacylation, decarboxylation, dihydroxylation, demethylation, dehalogenation, and conjugated hydrolysis reactions.[18] In a study of 271 oral pharmacological agents, it was found that many drugs were directly modified by the human microbiome.[31] Pharmacomicrobiomics is a new discipline to study how gut microbiota influence drug metabolism.[32] In other words, the microbiome is now viewed as an

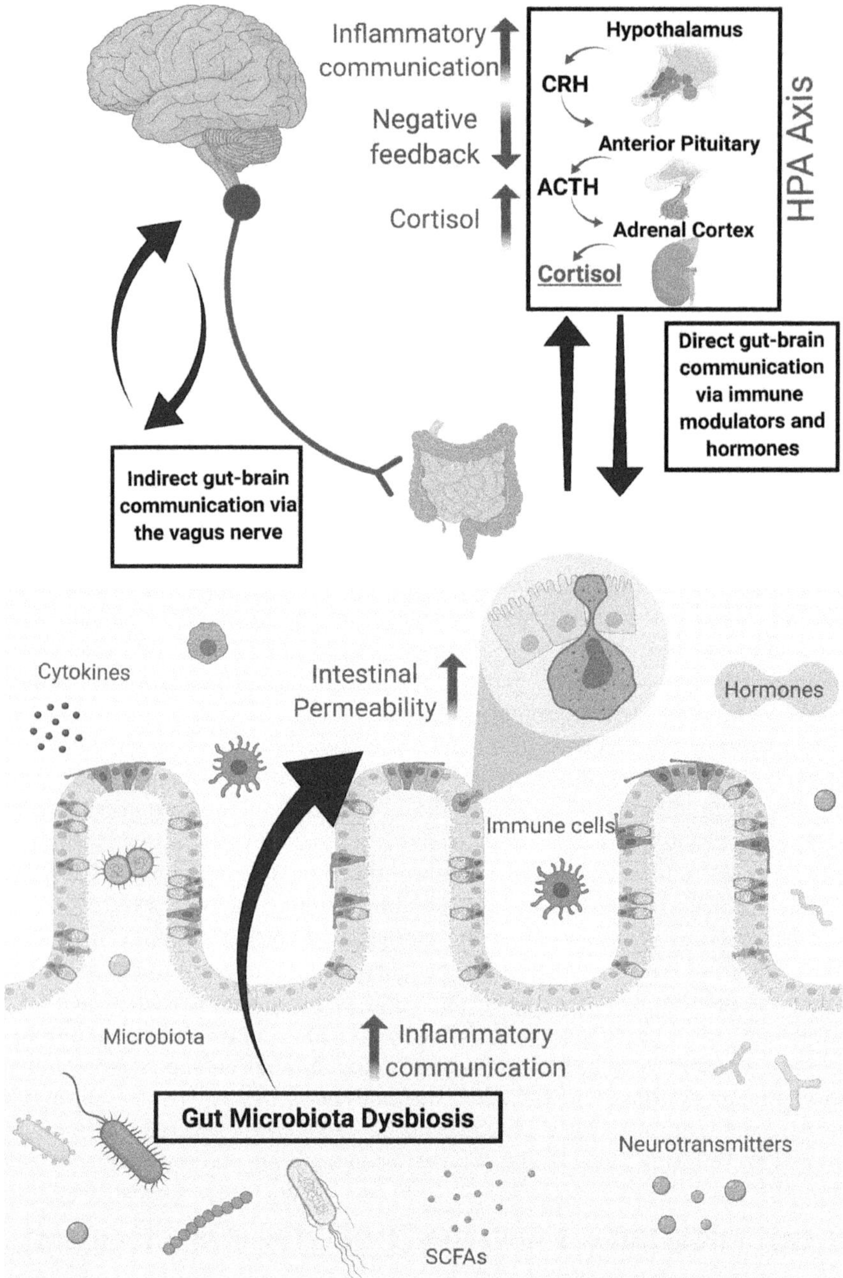

FIGURE 10.1 There are multiple pathways of communication between the microbiota located within the gastrointestinal tract and the brain. This often-overlooked influence could have implications for the etiology of anxiety disorders, for the efficacy and adverse effects of anxiolytic drug therapy (e.g., the benzodiazepines and "Z" drugs), and the nature and duration of withdrawal following the cessation of anxiolytic therapy.

active agent in drug metabolism rather than a bystander.[18] Acetaminophen, digoxin, metformin, cancer-treating drugs, and others have been shown to be affected by the microbiome and, in turn, to exert an influence on the gut microbiome.[33–35] It is likely that many other drugs are similarly affected.

Important in benzodiazepine metabolism is the nitroreductase activity of the microbiome. Nitroreductase activity involves reducing nitro-groups to amines. In a study of rats, the reductive metabolism of nitro-groups resulted in teratogenicity. Pregnant Sprague-Dawley rats were administered 300 mg/kg of nitrazepam, which led to the production of the metabolites 7-aminonitrazepam and 7-acetylaminonitrazepam. About 30% of the administered dose was excreted in the urine and feces. Rats were then administered antibiotics to deplete the microbiome and continued to receive nitrazepam, but now only 2% of the metabolites were excreted in urine and feces; that is, antibiotics suppressed the nitroreductase activity of the microbiome.[36] Incubation of clonazepam with rat intestinal lumen contents showed that the clonazepam was reduced almost entirely to the metabolite 7-aminoclonazepam. The hepatic microsomes of rats likewise could reduce clonazepam, but only under anaerobic conditions. Phenobarbital could induce aerobic and anaerobic microsomal metabolism in rats. Clonazepam metabolism in this murine study was mainly reductive in the presence of gut flora, but oxidative in the absence of gut flora (achieved via antibiotic pretreatment).[37]

Oral bromazepam has relatively low bioavailability in humans. In a study of four benzodiazepines (bromazepam, diazepam, desmethyldiazepam, and 7-bromo-1,3-dihydro-4-phenyl-2*H*-1,4-benzodiazpein-2(I*H*)-one), these agents were dissolved in a solution containing human fecal matter in order to investigate how the agents were affected. It was found that for all four agents, 80% degraded under both aerobic and anaerobic conditions.[38] Bromazepam was evaluated in a fecal medium made up of rat homogenate of liver and intestinal epithelial cells, and 70%–90% of the bromazepam remained in the medium after incubation for 5 h at 37°C. Thus, degradation of bromazepam by the microbiome must be considered a factor in benzodiazepine metabolism.[38]

ENDOZEPINES AND THE MICROBIOME

Of great interest are the so-called endozepines, recently elucidated endogenous benzodiazepines.[9] Mammals produce in their brain endogenous ligands that bear a strong chemical similarity to diazepam.[39] Certain microbes seem to be able to modulate the central nervous system in such a way that precursors of these endozepines are created.[40] Actually, it may go beyond that. More recent studies suggest that the human microbiome is able to synthesize numerous endogenous compounds that have a benzodiazepine-like core structure.[9] Observations have found that while some people seem to produce excessive amounts of endozepines, others produce too few, likely because of variations in the microbiome.[9] The lack of endogenous benzodiazepines may explain why certain people find exogenous benzodiazepine therapy helpful, even for prolonged courses of treatment.

By the same token, it has been observed that microbiota of the gut can produce metabolites with structures that resemble benzodiazepine molecules.[40–42] These metabolites may be detrimental to the host. In fulminant hepatic failure, benzodiazepine receptor ligands created in the gut microbiome may be contributors to encephalopathy.[40] Once in the brain, these endozepines may enhance the inhibitory quality of GABA receptors and in that way contribute to the disease.[40]

Pyrrolobenzodiazepines such as anthramycin can also be created in the gut microbiota and not only resemble benzodiazepines in chemical structure but may possess antibiotic and antineoplastic properties as well.[43] A pyrrolobenzodiazepine is a sequence-selective DNA-alkylating agent with antineoplastic properties that can be synthetically or naturally produced.[42]

This has led to an interesting observation. Patients with anxiety disorders may be suffering from problems originating in their gut microbiome rather than in the brain. In other words, people suffering from anxiety may be unable to produce sufficient endozepines and may thus benefit from exogenous benzodiazepines. This finding is supported by the fact that anxiety disorders and GI disorders have statistical correlations.[44] If this is borne out as a causal relationship—namely, that the gut microbiome may be at the root of many anxiety disorders—it may be possible to provide a therapeutic support of anxious patients by treating their gut dysbiosis rather than prescribing exogenous benzodiazepines or other psychiatric drugs.[9]

The association between GI complaints and mental health disorders deserves greater scrutiny.[45] In a study of 18,571 patients who underwent a structured interview, 6%–25% of them had unexplained GI symptoms.[44] Comparing patients with GI symptoms to others without such symptoms, those with GI symptoms had a higher lifetime rate of major depression (7.5% vs. 2.9%), panic disorder (2.5% vs. 0.7%), and agoraphobia (10.0% vs. 3.6%).[44] In patients with two GI symptoms, there were even higher rates of these mental health conditions (13.4% depression, 5.2% panic disorder, 17.8% agoraphobia). People with GI symptoms also have higher rates of anxiety and affective disorders.[44] Likewise, it has been shown that irritable bowel syndrome (IBS) patients with moderate to severe visceral pain may derive an analgesic benefit from benzodiazepines or other anxiolytic therapies, but do not get effective pain control from acetaminophen, opioids, or nonsteroidal anti-inflammatory drugs.[46] While the pathogenesis of IBS remains to be elucidated, intriguing differences in the microbiome of IBS and control patients have been discovered, suggesting that the microbiome may play an important role in the pathogenesis of IBS. Further supporting the link between the GI tract and the brain, a psychiatric comorbidity is present in up to 80% of IBS patients.[47]

THE GUT MICROBIOME, INSOMNIA, AND BENZODIAZEPINES

Certain benzodiazepines are indicated for the short-term relief of insomnia, a prevalent and somewhat heterogenous condition. The gut microbiome has been implicated in sleep patterns as well as mood disorders with the suggestion that the microbiota of the GI tract play a role in regulating circadian rhythm.[26] The majority of the one-celled organisms in the gut show diurnal fluctuations in activity.[48] Sleep deprivation and changes in sleep schedule have been shown to affect the gut microbiome.[49–52] Murine studies demonstrate that altering the sleep architecture of mice changes the diversity of their microbiome.[53] Likewise, a disruption of the host's circadian rhythm in humans alters their microbiome.[54,55]

Anxiety is associated with insomnia, which, in turn, appears to be related to gut microbiome dysbiosis.[56,57] The association among the microbiome, sleep disorders, and benzodiazepines is intriguing and suggests that insomnia may be related to the imbalances in the microbiome.

WHAT IS THE ROLE OF THE MICROBIOME IN BENZODIAZEPINE SIDE EFFECTS AND DEPENDENCE?

An important consideration in studying the microbiome is the variations in microbiota both among individuals and across community populations.[18] Since genetics, environment, diet, and other drug use can affect the microbiome, it is fair to say that the microbiome is unique to each individual. Furthermore, the microbiome is dynamic, changing over time. This has important ramifications for drug therapy, in that it may explain why certain patients are more responsive to a particular pharmacological treatment than other, similar patients with the same indications. The molecular mechanisms underlying the microbiome are not elucidated. Using high-throughput genetics and mass spectrometry, microbial gene products that might play a role in drug metabolism were identified. These microbiome-encoded enzymes play a role in drug metabolism that remain unclear, as awareness of the links between the genes of the microbiota and metabolic processes is only coming to the surface.[31] In the human GI tract, the microbiota is able to encode 150 times more genes than the human genome, exposing the drug to a diverse and extensive group of enzymes.[31] While this relatively new field of study offers tremendous potential for drug development and modern medicine, currently there is much that remains unknown, particularly in regard to the systematic understanding of the role of the microbiome in drug metabolism and the effect of the microbiome on specific agents, such as benzodiazepines.

An important consideration in studying the microbiome is the variations in microbiota both among individuals and across community populations.[18] Since genetics, environment, diet, and other drug use can affect the microbiome, it is fair to say that the microbiome is unique to each individual and dynamic, changing over time. This has important ramifications for drug therapy, in that it may explain why certain patients are more responsive to a particular pharmacological treatment than other, similar patients with the same indications. This may explain the variability among patients who take benzodiazepines. Some tolerate them well, others become dependent on them, some have side effects, and when benzodiazepines are discontinued, withdrawal symptoms vary widely among patients and are severe in some and virtually nonexistent in others.[58]

Furthermore, since the microbiota of the gut are known to produce endozepines and since the mind–gut axis has clear associations with symptoms such as anxiety, it would appear that the benzodiazepines and endozepines play prominent roles in the microbiome ecology. The challenge is that much needs to be learned about microbiota. While it is understood that the gut microbiota communicate with other organs and this likely involves neuropeptides, much about these communication avenues remains to be elucidated.[43]

FUTURE DIRECTIONS

Investigations into the microbiome and its role in drug metabolism remain in their infancy, and so far, the emphasis has been focused on the bacterial colonies of the microbiota. Although bacteria compose a large proportion of the microbiota, the influence of viruses and fungi has been largely ignored.[59] Thus, the research into the gut microbiome and its role in drug metabolism is clearly still in its infancy.

Furthermore, it is known that there are microbiome variations among individuals, within families, in ethnic groups, and in communities. The extent of these variations remains to be explored as do their potential ramifications of such variation on drug metabolism, immunity, and overall health. The gut microbiome may offer an explanation as to why certain patients are more responsive to particular drugs than others. In the case of benzodiazepines, the microbiome may explain the vast differences in side effects, development of dependence and tolerance, and withdrawal symptoms observed among patients taking these drugs, particularly those with prolonged exposure.

Finally, drug developers and research scientists have previously underestimated the gut microbiome in drug development. Clinical studies evaluate drug pharmacokinetics, dosing, effectiveness, and safety in patients with no regard to their gut microbiome. For benzodiazepine patients, the health and homeostasis of the gut microbiome may play a large role in terms of benzodiazepine metabolism. It is not yet clear where greater elucidation of the microbiome may take us—it may be that one day nonresponders to certain important drugs may be converted to responders through changes in the GI microbiota. In the case of benzodiazepines, it may be that endogenous benzodiazepines may be promoted as a natural byproduct of the microbiome rather than substituted with exogenous benzodiazepines.

CONCLUSION

The role of the microbiome in benzodiazepine metabolism is an important and intriguing area of study. While the gut microbiota are clearly involved in the metabolism of numerous pharmacological agents, the gut microbiome is also able to produce endozepines or endogenous molecules with a striking chemical similarity to benzodiazepines. Furthermore, GI disorders have been associated with anxiety, depression, and insomnia. Thus, it may be that certain individuals are able to reliably produce sufficient endozepines to regulate anxiety, while others require exogenous supplementation to provide anxiolysis. If this is indeed the case, future drug development or therapeutic innovations may be able to target microbiome dysbiosis to resolve anxiety and insomnia.

REFERENCES

1. Distrutti E, Monaldi L, Ricci P, Fiorucci S. Gut microbiota role in irritable bowel syndrome: new therapeutic strategies. *World J Gastroenterol.* 2016;22(7):2219–2241.
2. Hooper LV, Littman DR, Macpherson AJ. Interactions between the microbiota and the immune system. *Science.* 2012;336(6086):1268–1273.
3. Shade A, Caporaso JG, Handelsman J, Knight R, Fierer N. A meta-analysis of changes in bacterial and archaeal communities with time. *ISME J.* 2013;7(8):1493–1506.

4. Martin FP, Dumas ME, Wang Y, et al. A top-down systems biology view of microbiome-mammalian metabolic interactions in a mouse model. *Mol Syst Biol.* 2007;3:112.
5. Dietert RR, Silbergeld EK. Biomarkers for the 21st century: listening to the microbiome. *Toxicol Sci.* 2015;144(2):208–216.
6. Klaassen CD, Cui JY. Review: mechanisms of how the intestinal microbiota alters the effects of drugs and bile acids. *Drug Metab Dispos.* 2015;43(10):1505–1521.
7. Klünemann M, Schmid M, Patil KR. Computational tools for modeling xenometabolism of the human gut microbiota. *Trends Biotechnol.* 2014;32(3):157–165.
8. Turnbaugh PJ, Ley RE, Hamady M, Fraser-Liggett CM, Knight R, Gordon JI. The human microbiome project. *Nature.* 2007;449(7164):804–810.
9. Skolnick SD, Greig NH. Microbes and monoamines: potential neuropsychiatric consequences of dysbiosis. *Trends Neurosci.* 2019;42(3):151–163.
10. Lu K, Cable PH, Abo RP, et al. Gut microbiome perturbations induced by bacterial infection affect arsenic biotransformation. *Chem Res Toxicol.* 2013;26(12):1893–1903.
11. Neuman H, Koren O. The pregnancy microbiome. *Nestle Nutr Inst Workshop Ser.* 2017; 88:1–9.
12. Maruvada P, Leone V, Kaplan LM, Chang EB. The human microbiome and obesity: moving beyond associations. *Cell Host Microbe.* 2017;22(5):589–599.
13. Rosenfeld CS. Microbiome disturbances and autism spectrum disorders. *Drug Metab Dispos.* 2015;43(10):1557–1571.
14. Wang L, Hao K, Yang T, Wang C. Role of the lung microbiome in the pathogenesis of chronic obstructive pulmonary disease. *Chin Med J (Engl).* 2017;130(17):2107–2111.
15. Picardo SL, Coburn B, Hansen AR. The microbiome and cancer for clinicians. *Crit Rev Oncol Hematol.* 2019;141:1–12.
16. Szopinska-Tokov J, Dam S, Naaijen J, et al. Investigating the gut microbiota composition of individuals with attention-deficit/hyperactivity disorder and association with symptoms. *Microorganisms.* 2020;8(3) 406.doi:10.3390/microorganisms8030406.
17. Jochemsen R, Breimer DD. Pharmacokinetics of benzodiazepines: metabolic pathways and plasma level profiles. *Curr Med Res Opin.* 1984;8(sup4):60–79.
18. Wilson ID, Nicholson JK. Gut microbiome interactions with drug metabolism, efficacy, and toxicity. *Transl Res.* 2017;179:204–222.
19. Eckburg PB, Bik EM, Bernstein CN, et al. Diversity of the human intestinal microbial flora. *Science.* 2005;308(5728):1635–1638.
20. Cryan JF, Dinan TG. Mind-altering microorganisms: the impact of the gut microbiota on brain and behaviour. *Nat Rev Neurosci.* 2012;13(10):701–712.
21. Moloney RD, Desbonnet L, Clarke G, Dinan TG, Cryan JF. The microbiome: stress, health and disease. *Mamm Genome.* 2014;25(1–2):49–74.
22. Diaz Heijtz R, Wang S, Anuar F, et al. Normal gut microbiota modulates brain development and behavior. *Proc Natl Acad Sci U S A.* 2011;108(7):3047–3052.
23. Macpherson AJ, Harris NL. Interactions between commensal intestinal bacteria and the immune system. *Nat Rev Immunol.* 2004;4(6):478–485.
24. Breit S, Kupferberg A, Rogler G, Hasler G. Vagus nerve as modulator of the brain-gut axis in psychiatric and inflammatory disorders. *Front Psychiatry.* 2018;9:44.
25. Feng Q, Chen WD, Wang YD. Gut microbiota: an integral moderator in health and disease. *Front Microbiol.* 2018;9:151.
26. Li Y, Hao Y, Fan F, Zhang B. The role of microbiome in insomnia, circadian disturbance and depression. *Front Psychiatry.* 2018;9:669–669.
27. Powley TL, Wang XY, Fox EA, Phillips RJ, Liu LW, Huizinga JD. Ultrastructural evidence for communication between intramuscular vagal mechanoreceptors and interstitial cells of Cajal in the rat fundus. *Neurogastroenterol Motil.* 2008; 20(1):69–79.

28. Barrett E, Ross RP, O'Toole PW, Fitzgerald GF, Stanton C. γ-Aminobutyric acid production by culturable bacteria from the human intestine. *J Appl Microbiol.* 2012; 113(2):411–417.
29. Bienenstock J, Kunze W, Forsythe P. Microbiota and the gut-brain axis. *Nutr Rev.* 2015;73 Suppl 1:28–31.
30. Bravo JA, Forsythe P, Chew MV, et al. Ingestion of lactobacillus strain regulates emotional behavior and central GABA receptor expression in a mouse via the vagus nerve. *Proc Natl Acad Sci U S A.* 2011;108(38):16050–16055.
31. Zimmermann M, Zimmermann-Kogadeeva M, Wegmann R, Goodman AL. Mapping human microbiome drug metabolism by gut bacteria and their genes. *Nature.* 2019; 570(7762):462–467.
32. ElRakaiby M, Dutilh BE, Rizkallah MR, Boleij A, Cole JN, Aziz RK. Pharmacomicrobiomics: the impact of human microbiome variations on systems pharmacology and personalized therapeutics. *OMICS.* 2014;18(7):402–414.
33. Alexander JL, Wilson ID, Teare J, Marchesi JR, Nicholson JK, Kinross JM. Gut microbiota modulation of chemotherapy efficacy and toxicity. *Nat Rev Gastroenterol Hepatol.* 2017;14(6):356–365.
34. Currò D. The role of gut microbiota in the modulation of drug action: a focus on some clinically significant issues. *Expert Rev Clin Pharmacol.* 2018;11(2):171–183.
35. Li H, He J, Jia W. The influence of gut microbiota on drug metabolism and toxicity. *Expert Opin Drug Metab Toxicol.* 2016;12(1):31–40.
36. Takeno S, Sakai T. Involvement of the intestinal microflora in nitrazepam-induced teratogenicity in rats and its relationship to nitroreduction. *Teratology.* 1991;44(2):209–214.
37. Elmer GW, Remmel RP. Role of the intestinal microflora in clonazepam metabolism in the rat. *Xenobiotica.* 1984;14(11):829–840.
38. Fujii J, Inotsume N, Nakano M. Degradation of bromazepam by the intestinal microflora. *Chem Pharm Bull (Tokyo).* 1987;35(10):4338–4341.
39. Basile AS, Hughes RD, Harrison PM, et al. Elevated brain concentrations of 1,4-benzodiazepines in fulminant hepatic failure. *N Engl J Med.* 1991;325(7):473–478.
40. Yurdaydin C, Walsh TJ, Engler HD, et al. Gut bacteria provide precursors of benzodiazepine receptor ligands in a rat model of hepatic encephalopathy. *Brain Res.* 1995;679(1):42–48.
41. Hu Y, Phelan V, Ntai I, Farnet CM, Zazopoulos E, Bachmann BO. Benzodiazepine biosynthesis in Streptomyces refuineus. *Chem Biol.* 2007;14(6):691–701.
42. Gerratana B. Biosynthesis, synthesis, and biological activities of pyrrolobenzodiazepines. *Med Res Rev.* 2012;32(2):254–293.
43. Holzer P, Farzi A. Neuropeptides and the microbiota-gut-brain axis. *Adv Exp Med Biol.* 2014;817:195–219.
44. Walker EA, Katon WJ, Jemelka RP, Roy-Bryne PP. Comorbidity of gastrointestinal complaints, depression, and anxiety in the Epidemiologic Catchment Area (ECA) Study. *Am J Med.* 1992;92(1a):26s–30s.
45. Mayer EA, Craske M, Naliboff BD. Depression, anxiety, and the gastrointestinal system. *J Clin Psychiatry.* 2001;62 Suppl 8:28–36; discussion 37.
46. Chen L, Ilham SJ, Feng B. Pharmacological approach for managing pain in irritable bowel syndrome: a review article. *Anesth Pain Med.* 2017;7(2):e42747.
47. Collins SM. A role for the gut microbiota in IBS. *Nat Rev Gastroenterol Hepatol.* 2014;11(8):497–505.
48. Thaiss CA, Zeevi D, Levy M, et al. Transkingdom control of microbiota diurnal oscillations promotes metabolic homeostasis. *Cell.* 2014;159(3):514–529.
49. Davies SK, Ang JE, Revell VL, et al. Effect of sleep deprivation on the human metabolome. *Proc Natl Acad Sci U S A.* 2014;111(29):10761–10766.

50. Kunze KN, Hanlon EC, Prachand VN, Brady MJ. Peripheral circadian misalignment: contributor to systemic insulin resistance and potential intervention to improve bariatric surgical outcomes. *Am J Physiol Regul Integr Comp Physiol.* 2016;311(3):R558–R563.
51. Johnston JD, Ordovás JM, Scheer FA, Turek FW. Circadian rhythms, metabolism, and chrononutrition in rodents and humans. *Adv Nutr.* 2016;7(2):399–406.
52. Wu T, Yang L, Jiang J, et al. Chronic glucocorticoid treatment induced circadian clock disorder leads to lipid metabolism and gut microbiota alterations in rats. *Life Sci.* 2018;192:173–182.
53. Thaiss CA, Levy M, Korem T, et al. Microbiota diurnal rhythmicity programs host transcriptome oscillations. *Cell.* 2016;167(6):1495–1510.e1412.
54. Reynolds AC, Paterson JL, Ferguson SA, Stanley D, Wright KP, Jr., Dawson D. The shift work and health research agenda: considering changes in gut microbiota as a pathway linking shift work, sleep loss and circadian misalignment, and metabolic disease. *Sleep Med Rev.* 2017;34:3–9.
55. Reynolds AC, Broussard J, Paterson JL, Wright KP, Jr., Ferguson SA. Sleepy, circadian disrupted and sick: could intestinal microbiota play an important role in shift worker health? *Mol Metab.* 2017;6(1):12–13.
56. Flowers SA, Ellingrod VL. The microbiome in mental health: potential contribution of gut microbiota in disease and pharmacotherapy management. *Pharmacotherapy.* 2015;35(10):910–916.
57. Burokas A, Arboleya S, Moloney RD, et al. Targeting the microbiota-gut-brain axis: prebiotics have anxiolytic and antidepressant-like effects and reverse the impact of chronic stress in mice. *Biol Psychiatry.* 2017;82(7):472–487.
58. Ashton H. The diagnosis and management of benzodiazepine dependence. *Curr Opin Psychiatry.* 2005;18(3):249–255.
59. Nichols RG, Peters JM, Patterson AD. Interplay between the host, the human microbiome, and drug metabolism. *Hum Genomics.* 2019;13(1):27–27.

Section III

Implications for Therapeutics

11 Anxiety, Sleep, and Benzodiazepines

John F. Peppin
Marian University College of Osteopathic Medicine (Adjunct)
Pikeville University College of Osteopathic Medicine

CONTENTS

INTRODUCTION

Anxiety and sleep disorders affect a significant portion of the world's population.[1] In the United States, the prevalence of anxiety disorders is 18%–33% and is higher in females than in males.[1–3] In 2016, annual cost of this set of disorders was 42.3–46.6 billion USD.[1,4] Two of the three most common sleep disorders—sleep apnea and restless legs syndrome (RLS)—have a prevalence of 1.2% and 0.4% from a 2010 study.[5] Insomnia is the most common sleep disorder, and 50–70 million individuals have this disorder in the United States.[6] Overall, the presence of anxiety and sleep disorders has a significant impact on the health of individuals. How to treat these disorders has been an ongoing challenge for medicine and society. Unfortunately, a vast majority of anxiety and sleep disorders are untreated *vide infra*.[7,8] Further, both of these disorders are regularly and commonly treated with benzodiazepines (BZDs).

Once the BZD receptor had been identified, the search for endogenous ligands, or endogenous BZD-mimicking, "endozepines," started.[9] General interest in these substances has increased over the past decade, although the search for specific endozepines has proven elusive.[10]

The impact of endozepines in the development and treatment of insomnia and anxiety disorders is currently unknown; the literature on this subject is scant at best. However, since BZDs are so ubiquitous, the potential for disruption of a therapeutic outcome on insomnia and anxiety disorders due to receptor changes and endozepines is an area that requires further intense study.

Interestingly, during the COVID pandemic, the prevalence of anxiety and insomnia has increased dramatically. Two reviews found both anxiety and insomnia to be increased above previous non-COVID levels to a prevalence of 31.9% and 23.7%, respectively.[11,12] These increases were related to the stress of COVID.

ANXIETY

Anxiety Disorders was described in the early Hippocratic texts and continued to be included in texts from Cicero and to the present day.[13] This disorder has been listed since 1952 in the first iteration of DSM (1952), where is was listed within the chapter "psychoneurotic disorders," whereas in 1968, the DSM-II listed anxious symptoms as "neuroses."[13,14] The recent DSM-5 has included the pathophysiology and genetics in the categorization of anxiety disorders.[13] There are five major types of these disorders, namely, generalized anxiety disorder (GAD), obsessive–compulsive disorder (OCD), panic disorder, social anxiety disorder, and phobias.[15] Depression disorders are the most prevalent psychiatric disorders after anxiety.[16] Of the anxiety disorders listed in DSM-5, phobias are the most common.[17] Undertreatment of anxiety disorders is common, both nationally and internationally. In a European study, only 20.6% of those studies with anxiety disorder received professional help.[18] The age of onset for the various anxiety disorders varies. For example, GAD can start later in life, yet after the age of 50 anxiety disorders, other than GAD, decrease in incidence.[3] The research on biomarkers continues and has been reviewed; however to date, no specific biomarkers have been identified for anxiety.[19,20]

Although screening for anxiety disorders is fairly common, there is little evidence to suggest that this changes outcomes in the long term.[21] It has been suggested that "changes in practice structure" need to accompany the use of screening tools for anxiety disorders.[22] Such screens, however, should lead the healthcare professional to discussions with the patient, and a thorough history and physical, before any treatment is suggested. Anxiety does not exist in a vacuum, and a "wholistic" approach should be inculcated in any screenings being done.

The use of nonpharmacologic therapies in the treatment of anxiety should be a regular consideration for physicians presented with the anxious patient. Numerous studies have shown the efficacy of these approaches. A recent meta-analysis of group psychotherapy showed the benefit of group psychotherapy in the setting of anxiety disorders.[23] "Faith-adapted" cognitive behavioral therapy (CBT) may "outperform" standard CBT, although there were substantial limitations in the methodology of the review of this approach.[24] Interestingly, adding regular physical exercise to an individual's daily routine may reduce or even "protect" against the development of anxiety symptoms.[25] Although, a 2018 meta-analysis found the effect to be equivocal.[26] Yoga may also help the symptoms of anxiety.[27] Relaxation therapy has been used in the setting of anxiety disorders, and a recent Korean meta-analysis showed the benefit of this approach.[28] Acupuncture and electroacupuncture have been shown in a 2018 meta-analysis to be the effective therapy for anxiety, although secondary analysis suggested the data was of low quality.[29,30] Nevertheless, nonpharmacologic options should always be considered, and many of the above options can certainly help in other disease states as well. How these nonpharmaceutical approaches involve endozepines or the GABA receptors is unknown.

Patients with many psychiatric disorders, e.g., anxiety, have a functional deficiency of monamines, especially norepinephrine, serotonin, and dopamine. These are released by the presynaptic neuron, cross the synaptic cleft, and are taken up through the receptors in the postsynaptic membrane, which results in actions and effects on mood. Excess serotonin is taken up, "reuptake," into the cell for reuse (Figure 11.1). The selective serotonin reuptake inhibitors (SSRIs) and serotonin–norepinephrine reuptake inhibitors (SNRIs) both inhibit the reuptake of these neurotransmitters, allowing them to increase in the synaptic cleft, and thereby, their effect is increased.[31,32]

Unfortunately, only 54% of patients show improvement with medications.[33] Since the improvement rate is so low, other causative theories have been proposed, e.g., neurotrophic and brain-derived neurotrophic factor (BDNF), neuroendocrine mechanisms, and the neuroimmune and cytokine hypothesis.[32] In general, however, most FDA-approved antidepressants are believed to act by increasing serotonin, and/or norepinephrine, and/or dopamine availability by reducing reuptake.

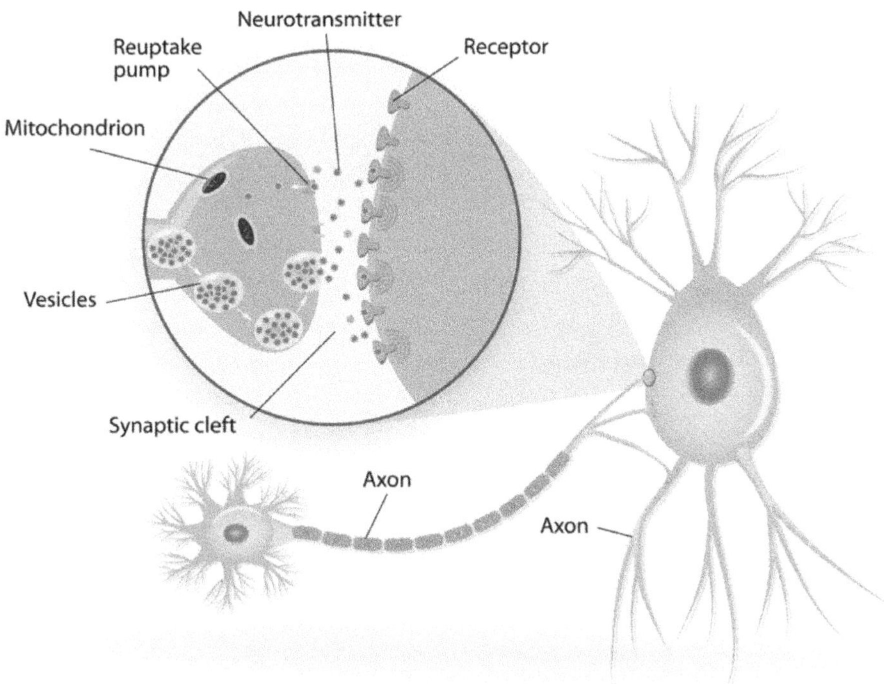

FIGURE 11.1 Illustration of a chemical synapse, showing the neurotransmitter reuptake sites. *Source*: purchased from iStock.

GABA is the most prevalent inhibitory neurotransmitter, and BZDs amplify its action. It is estimated that over 30% of neurons in the central nervous system have GABA as their primary neurotransmitter.[34] The GABA receptor is complex with multiple allosteric sites that allow a "level of inhibition of the neuron to be regulated with exquisite precision."[34] Multiple classes of drugs have been developed for treating a variety of disorders based on the GABA-A receptor, e.g., anticonvulsants, anxiolytics, and sedative–hypnotic drugs, including BZDs, barbiturates, neurosteroids, certain classes of general anesthetics, and alcohol, that bind to these multiple allosteric sites.[34] Binding of endogenous or exogenous BZDs to GABA-A receptors increases the amplitude and duration of the inhibitory postsynaptic currents.[35] However, GABA is not the only neurotransmitter that modulates anxiety responses in the amygdala (Figure 11.2), but others also play a role, e.g., serotonin, opioid peptides, endocannabinoids, neuropeptide Y, oxytocin, and corticotropin-releasing hormone.[34] In addition,

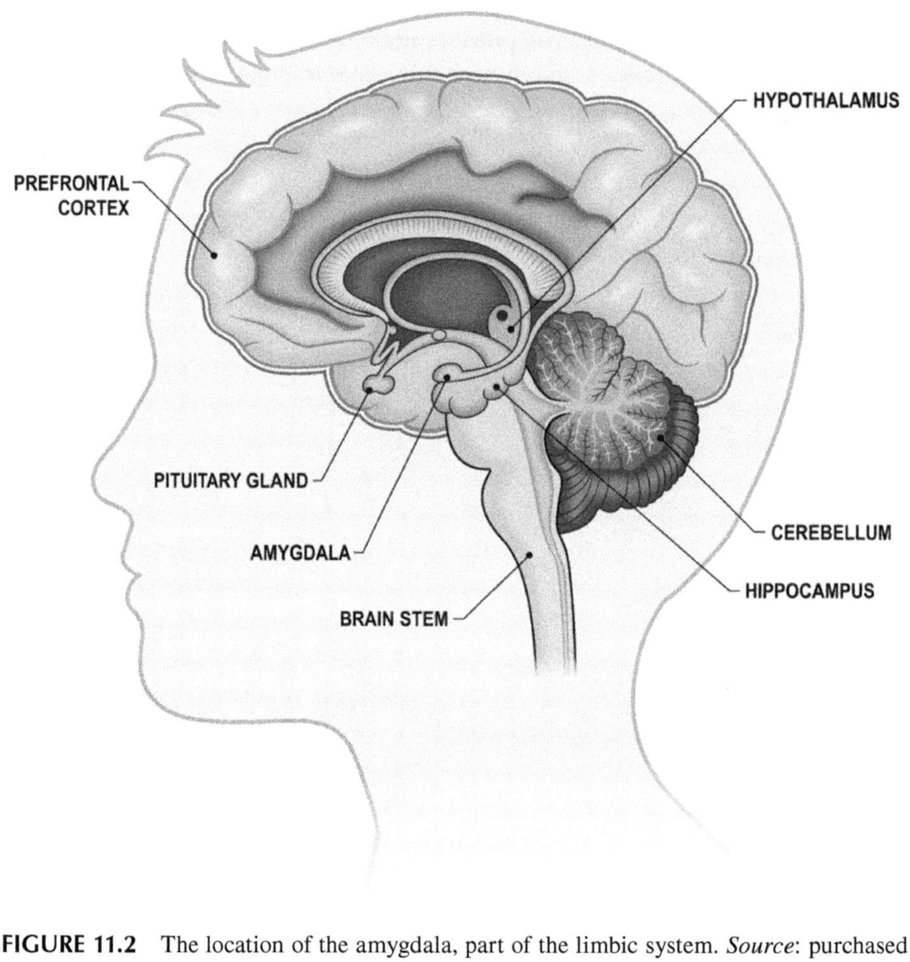

FIGURE 11.2 The location of the amygdala, part of the limbic system. *Source*: purchased from iStock.

neurosteroids binding to GABA receptors may also have anxiolytic and antidepressant actions, which have been described as "reminiscent of those elicited by barbiturates and BZDs."[34] Interestingly, neurosteroid synthesis is upregulated by SSRIs, which provides a link between anxiety treatment and the GABAergic system.[34,36] As with so many other chronic diseases, complexity is the rule, not the exception.

BZDs are one of the most commonly prescribed classes of medications currently in use. Further, compared to antidepressants as a class, BZDs have a high rate of long-term use.[37] Between 55 and 94% of patients with anxiety disorders are treated with BZDs.[38] These drugs are used in other areas of medicine, e.g., anesthesia, pain medicine, and sleep.[39]

The British Association for Psychopharmacology (BAP) has published guidelines and recommendations for the treatment of anxiety disorders.[22] When it comes to BZDs, BAP states, "BZDs will usually be reserved for the further treatment of patients who have not responded to at least three previous treatments."[22] The BAP review states that in the treatment of acute and chronic management of GAD, "substantial evidence for the efficacy of many antidepressant drugs" exists.[22] The World Federation of Biological Psychiatry (WFSBP) recommends the first-line therapy for anxiety gives both the strength of the category of evidence and also a recommendation grade.[40] The WFSBP recommends as first line for anxiety disorder, SSRIs, SNRIs, and pregabalin.[40] The BAP guidelines recommend "all" SSRIs, venlafaxine, reboxetine, "some TCAs," the MAOI phenelzine, and some BZDs (alprazolam, clonazepam, diazepam, and lorazepam), gabapentin, and valproate.[40] However, they state, "consider an SSRI for first-line pharmacological treatment."[22] Further, they recommend CBT, "over forms of psychological treatment." In some patients, the combination of an SSRI, specifically paroxetine, plus CBT may provide additional benefits.[41] As far as BZDs, although they give the data a "A" (full evidence from controlled studies), they give a two for recommendation; they recommend that BZDs be used, "only in the treatment of short-term distress."[40]

SLEEP DISORDERS

Approximately 34% of the population has transient insomnia, and about 40% of those develop chronic and persistent insomnia.[42]

Proper sleep quality is critical for health. There is a correlation between poor sleep quality, sleep disorders, and dementia.[43] Many other disease states are correlated with sleep disorders, and the cost to the health care system is $2,000 more per year, than a person without sleep disorders.[44] Sleep disorders impact the ability to fulfill daily tasks, memory, and learning.[45]

Diagnosing insomnia disorders can be made if the patient reports poor sleep onset and maintenance, as well as daytime somnolence, mood disturbances, and impaired attention. The diagnosis calls for these symptoms being present three nights a week and lasts for more than 3 months.[42] However, since insomnia does not exist in a vacuum, the evaluation of insomnia should be thorough and not cursory.

The treatment of insomnia covers a vast array of medications and nonpharmacologic therapies. The first step in treating insomnia is developing good sleep habits.

These include regular waking and sleep times, limiting time in bed to sleep and using bed for sleep and intimacy only. Avoiding caffeine after noon, limiting alcohol and avoiding daytime naps. If naps are taken, they should be brief, less than 30 min.[42] CBT has been shown to be either superior or similar to pharmacologic therapies.[46]

Insomnia is frequently accompanied with psychiatric disorders, e.g., anxiety, as well as other medical issues. These patients are currently being placed on antidepressants. Unfortunately, some classes of antidepressants can impact sleep quality.[47] SSRIs can cause insomnia, with an annual presence of about 17% of patients.[48] Since the occurrence of depression and anxiety is common in sleep disorders, insomnia can be made worse with the addition of antidepressants, and this should be included in the differential evaluation. Insomnia as a side effect of SSRIs is due to the activation and increase of serotonergic transmission.[47] However, drug classes with antihistaminergic activity can improve sleep, e.g., TCAs, mirtazapine, or mianserine. Further, those with strong serotonergic antagonistic action can also improve sleep, e.g., trazadone or nefazodone. For those who have sleep maintenance insomnia, longer-acting medications may be needed. And in those patients with other psychiatric disorders, such as anxiety or depression, mirtazapine or trazadone can be considered.[46] Nightly sleep maintenance insomnia may need nightly longer-acting medications, such as eszopiclone or suvorexant. Although cannabinoids have been suggested for use in insomnia, sleep problems increase the risk for relapse of a cannabis use disorder. Further, poor-sleep quality is a "hallmark" of withdrawal from cannabis.[49,50] The use of cannabinoids in the treatment of insomnia is still "in its infancy," but the myriad of side effects of cannabis, its potential for cannabis use disorder precludes it recommendation for use in insomnia.[49]

There are a multitude of guidelines, and a review of these is beyond the scope of this chapter. However, three will be briefly discussed. Two were done approximately a year apart, namely, The American College of Physicians (ACP) and the American Academy of Sleep Medicine (AASM), which developed guidelines that focus on chronic insomnia, and both used the GRADE approach for rating studies included in their recommendations.[51,52] The AASM guidelines focused on pharmaceutical therapies, whereas the ACP guidelines added a discussion of nonpharmaceutical options. For the BZDs, triazolam, and temazepam (other BZDs were not included), the AASM guidelines state, "we suggest that clinicians use triazolam (and temazepam) as a treatment for sleep onset insomnia."[52] Further, all of the AASM recommendations made are described as "weak" evidentiary support. In the ACP Guidelines however, "studies have shown that hypnotic drugs may be associated with infrequent but serious adverse effects, such as dementia, serious injury and fractures."[51] Further, "BZDs, *per se*, although widely used, were not addressed in this guideline because few studies met the inclusion criteria of the systematic review" and rated this as "insufficient evidence."[51] Their recommendation #1 "recommends that all adult patients receive cognitive behavioral therapy for insomnia as the initial treatment for chronic insomnia disorder."[51]

Guidelines developed by the European Sleep Research Society (ESRS) also support the use of CBT to treat chronic insomnia.[53] They also suggest that BZDs can be used "the aforementioned evidence suggests that BZDs and may be used in the short term if the first-line treatment (CBT-I) is ineffective or unavailable."[53] Their final recommendation, based on the evidence, was that BZDs and BZRAs are not recommended in the longer-term treatment of insomnia. This recommendation was rated as "strong evidence."[53] It is important to note that in Europe, eszopiclone, zolpidem, and suvorexant are not available. These three drugs take priority in the AASM recommendations, though they are rated as "weak" evidence.[52]

CONCLUSION

An increase in sleep disorders, as well as anxiety disorders, is seen in numerous sleep disorders, e.g., Parkinson's, kidney disease, thyroid disorders.[54,55] The above discussion should dissuade health care practitioners from taking a regimented approach to treatment. Patients should be evaluated individually and thoroughly, and therapeutic regimens should be applied on an individual basis. It is critical to understand the broad differential diagnoses when presented with an anxious patient or the patient with insomnia. The differential diagnoses can include other psychiatric disorders, such as major depression, but also physical syndromes and diseases such as hyperthyroidism and coronary artery disease.[3] All of these need to be considered before a clear therapeutic regimen can be designed. Anxiety and insomnia do not exist in a vacuum; unfortunately, they are frequently treated in this manner.

The use of BZDs internationally is ubiquitous. Although BZDs have a place in medicine, that place is narrow and this class of drug is currently overused.[10] Unfortunately, there are no focused guidelines for the specific use of and cessation of BZDs.[56] With the common prescribing of this class of drugs for everything from pain to panic attack, research does not support the use of BZDs in acute or chronic pain, neither for the long-term treatment of anxiety, nor as a first-line therapy for insomnia.[39] A recent ruling by the FDA illustrates the above comments. In October 2020, the FDA required a boxed warning for all BZDs.[57] The FDA's rationale was that "the current prescribing information for BZDs does not provide adequate warnings about these serious risks and harms associated with these medicines."[57]

In designing a therapeutic regimen for a given patient, both nonpharmacologic and pharmacologic approaches should be considered concomitantly. In mild or transient anxiety, nonpharmacologic therapies can be a better option than pharmaceutical options.[3] Combined therapy may reduce the amount of medication load, and have other benefits as well; e.g., exercise can benefit multiple systems in the body. In the United States, other pharmacologic options are available, and CBT, despite being effective, is underused for both of these conditions.

In conclusion, the contribution of endozepines in the development and treatment of anxiety and/or insomnia is currently unknown. However, clearly GABA-A receptors have an impact in the treatment of both of these disorders. Further research needs to be done in this area, so that an understanding of the impact of endogenous BZD substances can contribute to our treatment of these disorders.

REFERENCES

1. Remes O, Brayne C, van der Linde R, Lafortune L. A systematic review of reviews on the prevalence of anxiety disorders in adult populations. *Brain Behav.* 2016;6(7):e00497. doi:10.1002/brb3.497
2. Kessler RC, Berglund P, Demler O, Jin R, Merikangas KR, Walters EE. Lifetime prevalence and age-of-onset distributions of DSM-IV disorders in the National Comorbidity Survey Replication. *Arch Gen Psychiatry.* 2005;62(6):593–602. doi:10.1001/archpsyc.62.6.593
3. Bandelow B, Michaelis S. Epidemiology of anxiety disorders in the 21st century. *Dialogues Clin Neurosci.* 2015;17(3):327–35.
4. Devane CL, Chiao E, Franklin M, Kruep EJ. Anxiety disorders in the 21st century: status, challenges, opportunities, and comorbidity with depression. *Am J Manag Care.* 2005;11(12 Suppl):S344–53.
5. Ram S, Seirawan H, Kumar SK, Clark GT. Prevalence and impact of sleep disorders and sleep habits in the United States. *Sleep Breath.* 2010;14(1):63–70. doi:10.1007/s11325-009-0281-3
6. Association AS. Sleep and Sleep Disorder Statistics. Accessed November 30, 2020, 2020. https://www.sleepassociation.org/about-sleep/sleep-statistics/#:~:text=Sleep%20Disorder%20Statistics%3A,once%20in%20the%20preceding%20month.
7. Sartorius N, Ustün TB, Lecrubier Y, Wittchen HU. Depression comorbid with anxiety: results from the WHO study on psychological disorders in primary health care. *Br J Psychiatry Suppl.* 1996;30:38–43.
8. Park SC, Kim JM, Jun TY, et al. Prevalence and clinical correlates of insomnia in depressive disorders: the CRESCEND study. *Psychiatry Investig.* 2013;10(4):373–81. doi:10.4306/pi.2013.10.4.373
9. Christian CA, Herbert AG, Holt RL, et al. Endogenous positive allosteric modulation of GABA(A) receptors by diazepam binding inhibitor. *Neuron.* 2013;78(6):1063–74. doi:10.1016/j.neuron.2013.04.026
10. Peppin J, Pergolizzi J, Raffa R, Wright S. *The Benzodiazepines Crisis: The Ramifications of an Overused Drug Class.* First ed. Oxford University Press; 2020: 288.
11. Salari N, Hosseinian-Far A, Jalali R, et al. Prevalence of stress, anxiety, depression among the general population during the COVID-19 pandemic: a systematic review and meta-analysis. *Global Health.* 2020;16(1):57. doi:10.1186/s12992-020-00589-w
12. Cénat J, Blais-Rochette C, Kokou-Kpolou C, Noorishad P, Mukunzi J, McIntee S. Prevalence of symptoms of depression, anxiety, insomnia, posttraumatic stress disorder, and psychological distress among populations affected by the COVID-19 pandemic: a systematic review and meta-analysis. *Psychiatry Res.* 2021;(113599). doi:10.1016/j.psychres.2020.113599.
13. Crocq MA. A history of anxiety: from Hippocrates to DSM. *Dialogues Clin Neurosci.* 2015;17(3):319–25.
14. Association AP. *Diagnostic and Statistical Manual of Mental Disorders.* 2nd ed. American Psychiatric Association; 1968.
15. Services UDoHaH. What Are the Five Major Types of Anxiety Disorders? Updated February 12, 2014. Accessed December 3, 2020, 2020. https://www.hhs.gov/answers/mental-health-and-substance-abuse/what-are-the-five-major-types-of-anxiety-disorders/index.html
16. Kotiaho S, Korniloff K, Vanhala M, et al. Psychiatric diagnosis in primary care patients with increased depressive symptoms. *Nord J Psychiatry.* 2019;73(3):195–9. doi:10.1080/08039488.2019.1584240

17. Jacobi F, Höfler M, Strehle J, et al. Mental disorders in the general population: study on the health of adults in Germany and the additional module mental health (DEGS1-MH). *Der Nervenarzt.* 2014;85(1):77–87. doi:10.1007/s00115-013-3961-y.
18. Alonso J, Lépine JP, Committee EMS. Overview of key data from the European Study of the Epidemiology of Mental Disorders (ESEMeD). *J Clin Psychiatry.* 2007;68 Suppl 2:3–9.
19. Bandelow B, Baldwin D, Abelli M, et al. Biological markers for anxiety disorders, OCD and PTSD - a consensus statement. Part I: neuroimaging and genetics. *World J Biol Psychiatry.* 2016;17(5):321–65. doi:10.1080/15622975.2016.1181783
20. Bandelow B, Baldwin D, Abelli M, et al. Biological markers for anxiety disorders, OCD and PTSD: a consensus statement. Part II: neurochemistry, neurophysiology and neurocognition. *World J Biol Psychiatry.* 2017;18(3):162–214. doi:10.1080/15622975.2016.1190867
21. Christensen KS, Toft T, Frostholm L, Ornbøl E, Fink P, Olesen F. Screening for common mental disorders: who will benefit? Results from a randomised clinical trial. *Fam Pract.* 2005;22(4):428–34. doi:10.1093/fampra/cmi032
22. Baldwin DS, Anderson IM, Nutt DJ, et al. Evidence-based pharmacological treatment of anxiety disorders, post-traumatic stress disorder and obsessive-compulsive disorder: a revision of the 2005 guidelines from the British Association for Psychopharmacology. *J Psychopharmacol.* 2014;28(5):403–39. doi:10.1177/0269881114525674
23. Barkowski S, Schwartze D, Strauss B, Burlingame GM, Rosendahl J. Efficacy of group psychotherapy for anxiety disorders: a systematic review and meta-analysis. *Psychother Res.* 2020;30(8):965–82. doi:10.1080/10503307.2020.1729440
24. Anderson N, Heywood-Everett S, Siddiqi N, Wright J, Meredith J, McMillan D. Faith-adapted psychological therapies for depression and anxiety: systematic review and meta-analysis. *J Affect Disord.* 2015;176:183–96. doi:10.1016/j.jad.2015.01.019
25. McDowell CP, Dishman RK, Gordon BR, Herring MP. Physical activity and anxiety: a systematic review and meta-analysis of prospective cohort studies. *Am J Prev Med.* 2019;57(4):545–56. doi:10.1016/j.amepre.2019.05.012
26. Stonerock GL, Hoffman BM, Smith PJ, Blumenthal JA. Exercise as treatment for anxiety: systematic review and analysis. *Ann Behav Med.* 2015;49(4):542–56. doi:10.1007/s12160-014-9685-9
27. Cramer H, Lauche R, Anheyer D, et al. Yoga for anxiety: a systematic review and meta-analysis of randomized controlled trials. *Depress Anxiety.* 2018;35(9):830–43. doi:10.1002/da.22762
28. Kim HS, Kim EJ. Effects of relaxation therapy on anxiety disorders: a systematic review and meta-analysis. *Arch Psychiatr Nurs.* 2018;32(2):278–84. doi:10.1016/j.apnu.2017.11.015
29. Amorim D, Amado J, Brito I, et al. Acupuncture and electroacupuncture for anxiety disorders: a systematic review of the clinical research. *Complement Ther Clin Pract.* 2018;31:31–7. doi:10.1016/j.ctcp.2018.01.008
30. Li M, Xing X, Yao L, et al. Acupuncture for treatment of anxiety, an overview of systematic reviews. *Complement Ther Med.* 2019;43:247–52. doi:10.1016/j.ctim.2019.02.013
31. Jeon SW, Kim YK. Molecular neurobiology and promising new treatment in depression. *Int J Mol Sci.* 2016;17(3):381. doi:10.3390/ijms17030381
32. Harmer CJ, Duman RS, Cowen PJ. How do antidepressants work? New perspectives for refining future treatment approaches. *Lancet Psychiatry.* 2017;4(5):409–18. doi:10.1016/S2215-0366(17)30015-9
33. Cuijpers P, Stringaris A, Wolpert M. Treatment outcomes for depression: challenges and opportunities. *Lancet Psychiatry.* 2020;7(11):925–7. doi:10.1016/S2215-0366(20)30036-5

34. Nuss P. Anxiety disorders and GABA neurotransmission: a disturbance of modulation. *Neuropsychiatr Dis Treat.* 2015;11:165–75. doi:10.2147/NDT.S58841
35. Möhler H. Endogenous benzodiazepine site peptide ligands operating bidirectionally in vivo in neurogenesis and thalamic oscillations. *Neurochem Res.* 2014;39(6):1032–6. doi:10.1007/s11064-014-1303-5
36. Bloom FE, Iversen LL. Localizing 3H-GABA in nerve terminals of rat cerebral cortex by electron microscopic autoradiography. *Nature.* 1971;229(5287):628–30. doi:10.1038/229628a0
37. Starcevic V. The reappraisal of benzodiazepines in the treatment of anxiety and related disorders. *Expert Rev Neurother.* 2014;14(11):1275–86. doi:10.1586/14737175.2014.963057
38. Stahl SM. Don't ask, don't tell, but benzodiazepines are still the leading treatments for anxiety disorder. *J Clin Psychiatry.* 2002;63(9):756–7. doi:10.4088/jcp.v63n0901
39. Peppin J, Wright S. Benzodiazepines and Pain Management. In: Peppin J, Pergolizzi J, Raffa R, Wright S, eds. *The Benzodiazepine Crisis: The Ramifications of an Over-Used Drug Class.* First ed. Oxford University Press; 2021:149–59:chap 9.
40. Bandelow B, Sher L, Bunevicius R, et al. Guidelines for the pharmacological treatment of anxiety disorders, obsessive-compulsive disorder and posttraumatic stress disorder in primary care. *Int J Psychiatry Clin Pract.* 2012;16(2):77–84. doi:10.3109/13651501.2012.667114
41. Kampman M, Keijsers GP, Hoogduin CA, Hendriks GJ. A randomized, double-blind, placebo-controlled study of the effects of adjunctive paroxetine in panic disorder patients unsuccessfully treated with cognitive-behavioral therapy alone. *J Clin Psychiatry.* 2002;63(9):772–7. doi:10.4088/jcp.v63n0904
42. Pavlova M, Latreille V. Sleep disorders. *Am J Med.* 2019;132(3):292–9. doi:10.1016/j.amjmed.2018.09.021
43. Wennberg AMV, Wu MN, Rosenberg PB, Spira AP. Sleep disturbance, cognitive decline, and dementia: a review. *Semin Neurol.* 2017;37(4):395–406. doi:10.1055/s-0037-1604351
44. Xie Z, Chen F, Li WA, et al. A review of sleep disorders and melatonin. *Neurol Res.* 2017;39(6):559–65. doi:10.1080/01616412.2017.1315864
45. Amihăesei IC, Mungiu OC. Main neuroendocrine features and therapy in primary sleep troubles. *Rev Med Chir Soc Med Nat Iasi.* 2012;116(3): 862–6.
46. Jacobs GD, Pace-Schott EF, Stickgold R, Otto MW. Cognitive behavior therapy and pharmacotherapy for insomnia: a randomized controlled trial and direct comparison. *Arch Intern Med.* 2004;164(17):1888–96. doi:10.1001/archinte.164.17.1888
47. Wichniak A, Wierbicka A, Walecka M, Jernajczk W. Effects of antidepressants on sleep. *Curr Psychiatry Rep.* 2017;19(9):1–7. doi:10.1007/s11920-017-0816-4
48. Thompson C. Onset of action of antidepressants: results of different analyses. *Hum Psychopharmacol Clin Exp.* 2002;(S1):S27–32. doi:10.1002/hup.386
49. Babson KA, Boden MT, Harris AH, Stickle TR, Bonn-Miller MO. Poor sleep quality as a risk factor for lapse following a cannabis quit attempt. *J Subst Abuse Treat.* 2013;44(4):438–43. doi:10.1016/j.jsat.2012.08.224
50. Budney AJ, Moore BA, Vandrey RG, Hughes JR. The time course and significance of cannabis withdrawal. *J Abnorm Psychol.* 2003;112(3):393–402. doi:10.1037/0021-843x.112.3.393
51. Qaseem A, Kansagara D, Forciea MA, Cooke M, Denberg TD, Physicians CGCotACo. Management of chronic insomnia disorder in adults: a clinical practice guideline from the american college of physicians. *Ann Intern Med.* 2016;165(2):125–33. doi:10.7326/M15-2175
52. Sateia MJ, Buysse DJ, Krystal AD, Neubauer DN, Heald JL. Clinical practice guideline for the pharmacologic treatment of chronic insomnia in adults: an american academy of sleep medicine clinical practice guideline. *J Clin Sleep Med.* 2017;13(2):307–49. doi:10.5664/jcsm.6470

53. Riemann D, Baglioni C, Bassetti C, et al. European guideline for the diagnosis and treatment of insomnia. *J Sleep Res.* 2017;26(6):675–700. doi:10.1111/jsr.12594
54. Lajoie AC, Lafontaine AL, Kaminska M. The spectrum of sleep disorders in parkinson disease: a review. *Chest.* 2020;doi:10.1016/j.chest.2020.09.099
55. Nigam G, Camacho M, Chang ET, Riaz M. Exploring sleep disorders in patients with chronic kidney disease. *Nat Sci Sleep.* 2018;10:35–43. doi:10.2147/NSS.S125839
56. LeQuang J. In Search of Benzodiazepine Guidelines. In: Peppin J, Pergolizzi J, Raffa R, Wright S, eds. *The Benzodiazepines Crisis: An Overview of the Down-side of an Overused Drug Class.* first ed. Oxford University Press; 2020:216–36:chap 12.
57. (FDA) FaDA. FDA requiring Boxed Warning Updated to Improve Safe Use of Benzodiazepine Drug Class. FDA. Updated September 22, 2020. Accessed November 3, 2020, 2020. https://www.fda.gov/drugs/drug-safety-and-availability/fda-requiring-boxed-warning-updated-improve-safe-use-benzodiazepine-drug-class#:~:text=Facts%20about%20Benzodiazepines-,Benzodiazepines%20are%20a%20class%20of%20medicines%20approved%20to%20treat%20generalized,(see%20List%20of%20Benzodiazepines).

12 The Effects of Benzodiazepines on Memory

Jo Ann LeQuang
NEMA Research Inc.

CONTENTS

INTRODUCTION

Both acute and chronic benzodiazepine uses have been associated with memory deficits,[1] but the mechanisms of such memory dysfunction and the role of benzodiazepines remain to be elucidated. The potential role of endogenous benzodiazepines in memory has not been investigated with any rigor. Exogenous benzodiazepines are frequently prescribed for conditions such as insomnia and chronic pain, which in and of themselves are known to impair memory function.[2] Primary insomnia, an indication for benzodiazepine therapy, can negatively affect memory and decision-making processes even in people who take no benzodiazepines.[3] Cognitive deficits as well as adverse neuropsychological effects occur in some benzodiazepine patients, but not others, and those affected may find that these side effects persist long after benzodiazepine use has been discontinued.[1,4] It is not clear whether memory disruption or interference in benzodiazepine patients is the result of their underlying condition, the use of benzodiazepines, or an interplay between both factors. Thus, studies about memory in benzodiazepine patients have been confounded by the fact that many benzodiazepine patients have underlying conditions that may impair memory.

MEMORY AND MEMORY LOSS

Human memory is a vast, fascinating, and still largely unexplored field; what is known to date describes an exceedingly complex process that links together multiple discrete functions, specific aspects of many different cognitive processes, and a compartmentalized storage system with systems to move information from one compartment to another.[5] In simplistic terms, memory involves three specific steps in a linear progression: the acquisition of the memory (often called encoding), the consolidation of the memory, and ultimately its retrieval.[6] Memory appears to be substantially affected by many things such as motivation, attention level, emotional state, sensory perception, and so on.[6] The role of mood and its relationship with memory and learning is not well studied.[7] In addition, the three steps of encoding, consolidation, and retrieval can be tightly intertwined.

Memory is often described as short term or long term, which implies a temporal distinction that is less significant than the fact that these two main categories of memory are stored in different ways. Short-term memories are encoded in one of four ways (visually, acoustically, semantically, or by touch) and then are stored for <1 min in short-term memory. Short-term memories which are not rehearsed or reinforced in that brief window of time are promptly discarded, and the function of "active forgetting" appears to play a crucial, albeit not well-studied, role in memory. At any given time, there are about seven discrete items in short-term memory, which has an extremely limited storage capacity and functions primarily to discard extraneous information that it deems unhelpful. Some short-term memories migrate to long-term memory, which has more robust and longer storage, although long-term memories can also degrade over time.

Human memory can also be defined as explicit (e.g., one's address) or implicit (how to ride a bicycle) as well as episodic, sometimes called autobiographical (last weekend's activities), or semantic (the Fibonacci numbers). These various types of memory are processed in different ways, may be stored in different areas, are retrieved using different techniques, but at times may be required to work together in complex decision-making processes or abstract thought. For instance, the decision about buying a bicycle might include explicit memories (price comparisons), implicit memories (bicycling preferences), episodic memories (past cycling experiences) as well as reason and comparison along with emotion. It had long been theorized that exogenous benzodiazepines affect only explicit memory, but that may not be the case.[6]

Memories are initially labile and may remain extremely susceptible to alteration during the consolidation process, but with the passage of time, long-term memories stabilize. Consolidated memories are thought to be immune to amnesic agents but there is new evidence that suggests that even long-term memories can destabilize over time and resume their labile nature, and thus, even long-term memories may undergo "re-stabilization," during which time they may be susceptible to alteration as part of the consolidation process or be susceptible to amnesic agents. However, once a reconsolidated memory is stabilized again, it resists amnesics.[8]

An even more complex memory process is prospection which is a complex process involving episodic memory in the context of intentionality, allowing a person to be able to imagine a future or even fantastic event by relating it to past experience.[9] On the one hand, prospection can be a version of hypothetical thinking, such as weighing two different career options before making a decision. On the other hand, prospection can be used to allow a person to think about things that have never happened, such as an individual imaging what it would be like to be a movie star or to fight a dragon.[10] Such thought processes require memory to integrate with intention. Thus, memory deficits may not only impact memory retrieval, but they also have the potential to interfere with other thought processes.

In basic terms, memory retrieval involves recollecting a memory by locating it in the brain's storage systems and bringing it forward. While some memories are lost because they are not stored or have degraded with time or neglect, retrieval of extant memories is also susceptible to many errors, including reconstruction. For instance, an individual retrieving a long-term memory that has partially degraded may—without knowing it—fill in the gaps in the memory with likely but not necessarily accurate details. This explains why two people recollecting the same shared event may remember small aspects of the event differently, such as what one was wearing or the time of day.

THE EFFECT OF BENZODIAZEPINES ON MEMORY: WHAT IS KNOWN

Overall, benzodiazepines are held to be inhibitory modulators of memory processing.[4] The paradoxical ramifications of this are evident in the fact that exogenous benzodiazepines have been shown to both facilitate and impair memory depending on when they are administered. Benzodiazepine administration prior to the encoding of episodic memories is thought to impair memory function and, in that way, promote anterograde amnesia, while benzodiazepine administration immediately after memory encoding may facilitate memory.[11] This paradoxical effect is also observed with alcohol.[11]

In a study of volunteers giving a simple memory task on Day 1, subjects were brought in the next day and given a reminder presentation intended to labilize their memory. They were then administered 0.25 or 0.03 mg of clonazepam or placebo. Patients who took 0.25 mg of clonazepam exhibited an improvement in memory.[8] Yet, it has been speculated that short-acting benzodiazepines can interfere with reconsolidation of memories in a selective way via the gamma-aminobutyric acid (GABA)-A receptors, producing anterograde amnesia.[8] Thus, benzodiazepines may produce both memory-enhancing and memory-disrupting effects.

ANTEROGRADE AMNESIA

Benzodiazepines appear to impede memory acquisition (encoding) and blunt the brain's ability to form novel associations in memory.[6] Anterograde amnesia, which can be helpful for surgery patients, is often pharmacologically induced by acute administration of benzodiazepines.[12] The dose and half-life as well as the specific benzodiazepine agent used can be related to the degree of anterograde amnesia it produces, but it has been argued that all benzodiazepines are capable of producing at least some form of anterograde amnesia.[13,14] There is general, albeit not universal, consensus that benzodiazepines do not induce retrograde amnesia.[4,6]

Anterograde amnesia involves the disruption and interference in both the short- and long-term memory functions.[13] The anterograde amnesia associated with benzodiazepines is thought to be mediated by GABA-A receptor sites, suggesting that the anxiolytic effects of benzodiazepines may be functionally related to anterograde amnesia.[5]

GABA, MEMORY, AND BENZODIAZEPINES

GABA is a neurotransmitter with two receptors designated A and B. GABA-A receptors are involved mainly with fast inhibitory neurotransmission events and, as such, regulate anxiety, alertness, muscle tension, and memory.[4] GABA-A serves as a receptor for benzodiazepines and has at least 19 subunits, and it is thought that benzodiazepines working at GABA-A receptors may have cognitive and anxiolytic effects. The binding sites for benzodiazepines on the GABA-A receptors involve one of the three allosteric modulators: agonists, antagonists, and inverse agonists. Agonist activity is related to both anxiolysis and amnesia. Conversely, an inverse agonist activity can produce anxiety and enhance learning. Thus, when in good balance, GABA receptors appear to have helpful, bidirectional effects that promote sound-thought processes.[4] All of these GABA-A receptor subunits are pentameric membrane proteins that function, as a unit, as a GABA-gated chloride channel. Receptors often work together as two α-subunits paired with two β-subunits plus one γ-subunit to form one central ion channel. The binding site for benzodiazepine is typically located where the γ-subunit meets one of the α-units.[6]

As a class, benzodiazepines increase the inhibitory effects of GABA. GABA-ergic inhibition of hippocampal activity may inhibit thought processes. This may be beneficial and even therapeutic, such as when a person is experiencing intrusive thoughts, traumatic memories, or obsessive thought patterns, but it may also inhibit the memory of other, more benign things.[15] For instance, schizophrenia is characterized by hippocampal hyperactivity to the point that some patients experience hallucinations.[16] Abnormal hippocampal activity has also been documented in such conditions as post-traumatic stress disorder and major depression. It is likely that a dysfunction in GABA inhibition produces these symptoms, and the constellation of symptoms experienced by the patient may or may not be preceded by a stress-inducing or stress-triggering event.[15,17] GABA inhibition appears to be able to stop the memory retrieval process involving the hippocampus, but this is not entirely elucidated.[18] People living with type 2 diabetes often experience episodic memory deficits, and a study found such patients had abnormal GABA levels in their medial prefrontal cortex.[19] Anecdotally, people on long-term benzodiazepine therapy have been observed to exhibit deficiencies in episodic memory.

Benzodiazepine treatment might be effective in treating anxiety and other conditions precisely because it can pharmacologically interfere with the production of intrusive and overwhelming thoughts.[18] As an illustration, a disturbing thought arising in the prefrontal cortex will emerge unless the brain has sufficient GABA to inhibit it; benzodiazepines potentiate the effect of GABA, and this boost in GABA in the brain prevents the retrieval of that distressing idea or image. However, it remains to be elucidated as to whether benzodiazepines block other information and memories from being retrieved as well.[18]

THE PURPOSE OF FORGETTING: WHAT IS BEING LEARNED

Forgetfulness or the purposeful discarding of memories may be a crucial although poorly understood function of the human brain.[18] While short-term memories reside only briefly in memory and are routinely discarded all day long, long-term memories dwell in the hippocampus and link out into a neural network that strengthens itself the more frequently the memory is recalled. Frequent recall literally creates a physical form for the memory known as an engram. Engrams are enmeshed in a network of neurons, such that one individual neuron may link to multiple different engrams. Alpha-amino-3-hydroxy-5-methyl-4-isoxazolepropionic acid receptors (AMPA receptors) are ionotropic transmembrane receptors for glutamate that facilitate fast synaptic transmission across the neural network of the central nervous system. It is the AMPA receptors that keep memories intact, but AMPA receptors are notoriously unstable. They migrate, die off, and regenerate in such a way that this glue holding human memories intact is fundamentally unreliable.[18] Consistent and frequent recollection of memory can help improve the stability of AMPA receptors at the synapses, but there is evidence that suggests these AMPA receptors may be deactivated in order to promote forgetting as a necessary functional activity of the brain.[18] In other words, memory loss in certain contexts can be beneficial. Over the course of time, memories tend to be generalized around the key evident or message rather than the extraneous details. For example, a person may remember a small traffic accident from years ago without remembering what he was wearing at the time, the date, or even the name of the other party involved. This sort of purposeful culling of detail allows a clearer focus on the key aspects of the memory rather than getting mired in details.

Super autobiographical memory (HSAM) is a rare and intriguing condition, in which people remember virtually every intricate detail of all or most events they experience. They may be able to recollect what they were watching on TV on a specific night 10 years ago or repeat old conversations verbatim. However, people with HSAM are no more immune to false memories than people without HSAM,[20] so it appears that the fundamental memory processes are the same in individuals with or without HSAM, except that people with HSAM store more memories and have certain structural differences in their brain.[21] While HSAM is astonishing, it does not seem to predispose the individuals who have this condition to great accomplishments or ability. In fact, there is a tendency among people with HSAM to develop obsessive tendencies.[18] At the opposite end of the spectrum is severely deficient autobiographical memory (SDAM), which occurs when people cannot recall even the most important, dramatic, or vivid events of their life. It is interesting to note that SDAM can facilitate abstract thinking, problem solving, and certain creative processes.[18]

The ability to forget information may play an important role in many mental health conditions. For example, post-traumatic stress disorder could be described as a disorder in which horrifying events simply cannot be forgotten and intrude on daily life.[18] Anxiety may be the inability to discard persistent worrisome thoughts. Benzodiazepines might play a therapeutic role in such cases by facilitating the individual's ability to forget noxious memories. In contrast, Alzheimer's disease is currently being re-framed by some investigators as a neurological disorder in which the brain's ability to forget information occurs too randomly and rapidly.[18]

SLEEP, MEMORY, AND BENZODIAZEPINES: WHAT IS EMERGING

Sleep is involved in memory consolidation in ways that have yet to be fully elucidated. It is widely held that memory consolidation occurs during sleep, although there is some evidence that suggests memories may also be consolidated during wakefulness.[22] Normal sleep architecture is defined by four stages of non-rapid-eye-movement (non-REM) sleep and one stage of REM sleep. Non-REM sleep stages last about 90min each, but REM sleep durations vary with age and other factors. Infants, for example, get more REM sleep than older people. Each sleep stage is characterized by its own unique electroencephalographic oscillations, physiological responses, and dream activity. All sleep stages except non-REM stage 1 are thought to be involved in memory consolidation but it is not clear if certain sleep stages process specific types of memories.[22] It has been suggested that emotionally charged memories are more likely to be consolidated during periods of REM sleep than non-REM sleep, and that non-REM sleep stages consolidate more emotionally neutral memories.[22]

If sleep is needed to consolidate memories, then insomnia would likely cause memory problems, and indeed, insomnia is associated with numerous cognitive deficits, including memory loss.[2,3] Intriguingly, parasomnias, such as night terrors or sleep walking, are not associated with memory deficits.[23] Since benzodiazepines and related so-called Z-drugs are indicated for the treatment of insomnia, it is unclear if memory dysfunction in this population can be attributed all or in part to their underlying sleep disorder rather than benzodiazepine therapy.

It is known that the use of benzodiazepines disrupts sleep architecture in ways that vary by agent, dose, and product formulation.[24,25] Benzodiazepines as a class reduce three of the four stages of sleep associated with memory consolidation (REM, non-REM stage 3 and stage 4).[26,27] Furthermore, benzodiazepine use may affect the cyclic alternating patterns of restful sleep, that is, the microarchitecture of sleep.[28] It has been observed that benzodiazepine tapering or discontinuation often results in so-called REM rebound, where patients experience intense periods of increased REM sleep activity and vivid dreams.[27]

Benzodiazepines are indicated for short-term treatment of insomnia, but as insomnia is often a chronic condition, it is not unusual to find insomniacs on chronic, off-label benzodiazepine therapy. Memory impairment has been documented in insomniacs treated with long-term benzodiazepine therapy and may even be treatment-limiting. While a meta-analysis concluded that benzodiazepines increased sleep duration in patients with insomnia, long-term benzodiazepine use was associated with numerous side effects, including, but not limited to, memory loss.[29]

CLINICAL STUDIES AND SPECIAL POPULATIONS: WHAT IS BEING LEARNED

While memory impairment has long been recognized as a side effect of benzodiazepine use,[30] even to the point that benzodiazepines are administered to presurgical patients as amnesic agents, many people on long-term benzodiazepine therapy are unaware of any memory deficits, even when they are apparent to those around them.[12] Others on long-term benzodiazepine therapy may realize they are having memory problems but be too embarrassed to admit it or may attribute the memory loss to other things, such as advancing age.[12] A meta-analysis of long-term benzodiazepine patients found statistically significant deficits in working memory, memory processing, recent memory, and divided attention compared to controls.[1] High-dose benzodiazepine-dependent individuals may experience anterograde amnesia and disinhibition when taking short-acting benzodiazepines particularly in combination with alcohol. This can result in their finding out after the fact that they indulged in behaviors that embarrassed them or caused shame; many of them only became aware of the amnesia when peers told them about the things they had done while under the influence.[31]

In a study of 20 benzodiazepine patients at a rehabilitation center (aged ≥ 55 years, detoxified) matched with 20 alcohol-dependent patients in the same program and with 22 controls from the community, neuropsychological testing was carried out at 6–10 days after detoxification. The benzodiazepine patients had more difficulty with learning, short-term recall, and delayed recall than either the alcohol-dependent or control patients.[32] In another study of 33 older patients (mean age 73.6 years) in a rehabilitation program for benzodiazepine use disorder, it was found that patients had significant deficits in visual immediate recall, visual delayed recall, working memory, verbal learning, verbal fluency, speech processing, and sustained attention compared to control patients.[33] Memory impairments were also observed in community-dwelling older people on long-term benzodiazepine therapy compared to similar patients who did not take benzodiazepines.[34]

Benzodiazepine use was studied in selected special populations with respect to memory impairment.

Dementia Patients

Amyloid plaques are misfolded proteins that develop in neural cells and in the brain, which have been associated with Alzheimer's disease and other forms of neurocognitive impairment. The β-amyloid peptide has been extensively studied, and it is known that the Aβ peptides are small fragments of an amyloid precursor protein (APP). In order for the Aβ to be separated from the APP, the APP has to be cleaved by the β- and γ-secretase enzyme in sequence. The β-site amyloid precursor protein-cleaving enzyme 1 (BACE-1) has been a target of drug discovery for Alzheimer's disease therapeutics with the idea that inhibition of BACE-1 might inhibit Alzheimer's disease. Benzodiazepines are known to inhibit both BACE-1 and γ-secretase activity, which reduce the oligomers Aβ in the brain.[35] Reducing the aggregation of Aβ reduces the risk of developing Alzheimer's disease, but the disease is more complex.

Amyloid plaques in the brain are often surrounded by astrocytes that secrete GABA. In patients with early amyloid lesions, this can lead to an increase in GABA, which is potentiated by benzodiazepine use.[36] In patients with dementia, it has been speculated that early in the disease, compensatory neural networks develop in order to minimize the effect of early amyloid plaque on the brain, but exogenous benzodiazepines may lower the brain activity level to the point that it impedes this effort and, in that way, may accelerate cognitive decline.[36] Indeed, benzodiazepine use has been associated with elevated risk of development of Alzheimer's disease or exacerbation of preexisting Alzheimer's disease.[37]

Insomnia Patients

Studies of patients taking benzodiazepines for insomnia have produced conflicting results in terms of memory loss. In a study of older adults (49 with insomnia and 49 controls, mean age 69 years), cognitive performance was similar between those who slept well and those with sleep disorders, and both groups had similar working memory abilities.[38] However, another study comparing 64 geriatric adults without insomnia and 35 with insomnia found that those with insomnia had significant deficits in memory span, attention, estimating time, executive function, and the integration of visual and semantic dimensions.[39] In a study of women with nonmetastatic breast cancer, women with insomnia had significantly poorer scores on verbal episodic memory and executive function compared to similar patients without insomnia.[40]

Panic Disorder Patients

As mentioned earlier, many studies of neurocognitive deficits associated with long-term benzodiazepine use have been confounded by the fact that people taking benzodiazepines for chronic therapy often have underlying psychological or mental conditions which may adversely impact their memory. For example, a study of 20 panic disorder patients on benzodiazepine therapy matched to 20 panic disorder patients not taking benzodiazepines found that both groups had a similar degree of nonverbal memory impairment independent of benzodiazepine therapy.[41] This implies that memory deficits in panic disorder patients may be attributable to the underlying disease rather than benzodiazepine therapy.

People Living with HIV

Benzodiazepine use is prevalent among people living with HIV. In a study of this population ($n = 306$) taking chronic benzodiazepine therapy (median 24 months), the current use of a benzodiazepines could be statistically correlated with a higher probability of neurocognitive impairment (odds ratio [OR] 2.13, $p = 0.003$). The benzodiazepines diazepam, lorazepam, and alprazolam were most associated with neurocognitive impairment (OR > 2.31).[42]

People on Opioid Substitution Therapy

People with opioid-use disorder may be treated with opioid substitution therapy, typically using methadone, buprenorphine, or a combination product of buprenorphine plus naloxone, to facilitate recovery. In a memory study of such patients who were also taking benzodiazepines, a comparison was made among 13 methadone-maintained patients, 15 buprenorphine or buprenorphine plus naloxone patients, and 15 controls. Two tests were conducted, 1 about 2 months after admission into the opioid rehabilitation program and the other 6–9 months after admission. Both groups of opioid substitution therapy patients performed significantly worse in memory tests than control patients. This suggests that working memory is impaired in opioid substitution patients taking benzodiazepines.[43]

Schizophrenia Patients

In a study of 407 community-dwelling stabilized schizophrenia patients, long-term use of benzodiazepines was associated with impaired working memory as well as impaired attention (OR 0.60, 95% confidence interval, 0.42–0.86, $p = 0.005$) after adjusting for sociodemographic variables and characteristics of the illness.[44]

Suicidal Patients

A study evaluated 43 patients who attempted suicide by benzodiazepine overdose. Patients were admitted to the hospital and evaluated on the first and second day following admission. Patients had less immediate recall and less delayed recall in the first 24 h after admission compared to the first 48 h, when immediate and delayed recall improved somewhat. All patients had anterograde amnesia. This has implications for people with suicidal ideation who may overdose on benzodiazepines, because they may not remember what they had taken or their motivations.[45,46]

Women versus Men

In a study of 1,019 French salaried individuals followed for 10 years, 3.9% reported themselves as occasional and 7.5% as long-term users of benzodiazepines. Patients were recruited by age class and were 32, 42, 52, or 62 years at baseline. At baseline testing, in their respective age groups, women scored better on all cognitive tests than men. In memory tests of the individuals after baseline, women taking benzodiazepines exhibited more long-term memory impairment than men taking benzodiazepines. In addition, for both sexes, long-term benzodiazepine users had more memory impairment than occasional users.[47] This is one of the few studies that explored benzodiazepine memory impairment by sex, although cognitive differences have been well studied.

PERSISTENCE OF BENZODIAZEPINE-ASSOCIATED MEMORY LOSS

Cognitive effects associated with long-term benzodiazepine use may not abate upon discontinuation of benzodiazepine therapy.[30] Long-term benzodiazepine use may be associated with deficits in verbal memory, nonverbal memory, attention, and concentration that may persist even up to 6 months or more of benzodiazepine abstinence.[48] However, it is not clear whether these cognitive impairments were induced solely by benzodiazepines, exacerbated by them, or were the result of some underlying conditions.

CONCLUSIONS

Benzodiazepines can produce anterograde (but not retrograde) amnesia and are known to affect memory in other ways, likely through their potentiation of the GABA-A inhibitory pathways. However, many conditions that indicate the use of benzodiazepines, such as insomnia or anxiety, may produce cognitive impairment apart from benzodiazepine therapy. In fact, it is challenging to distinguish what memory loss in benzodiazepine patients is due to benzodiazepine use versus the underlying disorder, or if some interplay between them is responsible. It is alarming that memory deficits occurring with long-term benzodiazepine use do not immediately resolve upon drug discontinuation and may persist for prolonged periods. Greater elucidation of the various memory processes, including memory consolidation and recall, may help better clarify the issue. However, memory impairment and cognitive deficits must be considered when prescribing—and discontinuing—benzodiazepines.

REFERENCES

1. Crowe SF, Stranks EK. The residual medium and long-term cognitive effects of benzodiazepine use: an updated meta-analysis. *Arch Clin Neuropsychol* 2018;33(7):901–911.
2. Bothelius K, Hysing EB, Filén T, Lundeborg L, Gordh T. Insomnia-related memory impairment in individuals with very complex chronic pain. *Cogn Behav Neurol.* 2019;32(3):164–171.
3. Chunhua X, Jiacui D, Xue L, Kai W. Impaired emotional memory and decision-making following primary insomnia. *Medicine (Baltimore).* 2019;98(29):e16512.
4. Savić MM, Obradović DI, Ugresić ND, Bokonjić DR. Memory effects of benzodiazepines: memory stages and types versus binding-site subtypes. *Neural Plast.* 2005; 12(4):289–298.
5. Barbee JG. Memory, benzodiazepines, and anxiety: integration of theoretical and clinical perspectives. *J Clin Psychiatry.* 1993;54 Suppl:86–97; discussion 98–101.
6. Abel T, Lattal KM. Molecular mechanisms of memory acquisition, consolidation and retrieval. *Curr Opin Neurobiol.* 2001;11(2):180–187.
7. Curran HV. Benzodiazepines, memory and mood: a review. *Psychopharmacology.* 1991;105(1):1–8.
8. Rodríguez ML, Campos J, Forcato C, et al. Enhancing a declarative memory in humans: the effect of clonazepam on reconsolidation. *Neuropharmacology.* 2013;64:432–442.
9. Osman M. What are the essential cognitive requirements for prospection (thinking about the future)? *Front Psychol.* 2014;5:626.
10. Gilbert DT, Wilson TD. Prospection: experiencing the future. *Science.* 2007;317(5843): 1351–1354.

11. Doss MK, Weafer J, Ruiz NA, Gallo DA, De Wit H. Alcohol and pharmacologically similar sedatives impair encoding and facilitate consolidation of both recollection and familiarity in episodic memory. *Cogn Neurosci.* 2018;9(3–4):89–99.
12. Mejo SL. Anterograde amnesia linked to benzodiazepines. *Nurse Pract.* 1992;17(10): 44, 49–50.
13. Roth T, Roehrs T, Wittig R, Zorick F. Benzodiazepines and memory. *Br J Clin Pharmacol.* 1984;18(Suppl 1):45s–49s.
14. Moussaoui D. Benzodiazepines and memory. *Encephale.* 1986;12(6):315–319.
15. Schmitz TW, Correia MM, Ferreira CS, Prescot AP, Anderson MC. Hippocampal GABA enables inhibitory control over unwanted thoughts. *Nat Commun.* 2017;8(1):1311.
16. Heckers S, Konradi C. GABAergic mechanisms of hippocampal hyperactivity in schizophrenia. *Schizophr Res.* 2015;167(1–3):4–11.
17. Rayner G, Jackson G, Wilson S. Cognition-related brain networks underpin the symptoms of unipolar depression: evidence from a systematic review. *Neurosci Biobehav Rev.* 2016;61:53–65.
18. Gravitz L. The forgotten part of memory. *Nature.* 2019;571(7766):S12–S14.
19. Thielen JW, Gancheva S, Hong D, et al. Higher GABA concentration in the medial prefrontal cortex of Type 2 diabetes patients is associated with episodic memory dysfunction. *Hum Brain Mapp.* 2019;40(14):4287–4295.
20. Patihis L, Frenda SJ, LePort AKR, et al. False memories in highly superior autobiographical memory individuals. *Proc Natl Acad Sci U S A.* 2013;110(52):20947.
21. LePort AKR, Mattfeld AT, Dickinson-Anson H, et al. Behavioral and neuroanatomical investigation of highly superior autobiographical memory (HSAM). *Neurobiol Learn Mem.* 2012;98(1):78–92.
22. Stickgold R. Slee-dependent memory consolidation. *Nature.* 2005;427:1272–1278.
23. Cellini N. Memory consolidation in sleep disorders. *Sleep Med Rev.* 2017;35:101–112.
24. Tan X, Uchida S, Matsuura M, Nishihara K, Kojima T. Long-, intermediate- and short-acting benzodiazepine effects on human sleep EEG spectra. *Psychiatry Clin Neurosci.* 2003;57(1):97–104.
25. Tan X, Uchida S, Matsuura M, Nishihara K, Iguchi Y, Kojima T. Benzodiazepine effects on human sleep EEG spectra: a comparison of triazolam and flunitrazepam. *Life Sci.* 1998;63(8):675–684.
26. Roehrs T, Roth T. Drug-related sleep stage changes: functional significance and clinical relevance. *Sleep Med Clin.* 2010;5(4):559–570.
27. Pagel JF, Parnes BL. Medications for the treatment of sleep disorders: an overview. *Prim Care Companion J Clin Psychiatry.* 2001;3(3):118–125.
28. Manconi M, Ferri R, Miano S, et al. Sleep architecture in insomniacs with severe benzodiazepine abuse. *Clin Neurophysiol.* 2017;128(6):875–881.
29. Holbrook AM, Crowther R, Lotter A, Cheng C, King D. Meta-analysis of benzodiazepine use in the treatment of insomnia. *CMAJ.* 2000;162(2):225–233.
30. Ashton H. The diagnosis and management of benzodiazepine dependence. *Curr Opin Psychiatry.* 2005;18(3):249–255.
31. Liebrenz M, Gehring MT, Buadze A, Caflisch C. High-dose benzodiazepine dependence: a qualitative study of patients' perception on cessation and withdrawal. *BMC Psychiatry.* 2015;15:116.
32. Rummans TA, Davis LJ, Jr., Morse RM, Ivnik RJ. Learning and memory impairment in older, detoxified, benzodiazepine-dependent patients. *Mayo Clin Proc.* 1993; 68(8):731–737.
33. Ros-Cucurull E, Palma-Álvarez RF, García-Raboso E, et al. Benzodiazepine use disorder and cognitive impairment in older patients: a six-month-follow-up study in an outpatient unit in Barcelona. *J Stud Alcohol Drugs.* 2018;79(6):844–852.

34. Hanlon JT, Horner RD, Schmader KE, et al. Benzodiazepine use and cognitive function among community-dwelling elderly. *Clin Pharmacol Ther.* 1998;64(6):684–692.
35. Yan R. Physiological functions of the β-site amyloid precursor protein cleaving enzyme 1 and 2. *Front Mol Neurosci.* 2017;10(97). doi: 10.3389/fnmol.2017.00097.
36. Pariente A, de Gage SB, Moore N, Bégaud B. The benzodiazepine-dementia disorders link: current state of knowledge. *CNS Drugs.* 2016;30(1):1–7.
37. Billioti de Gage S, Moride Y, Ducruet T, et al. Benzodiazepine use and risk of Alzheimer's disease: case-control study. *BMJ (Clin Res Ed).* 2014;349:g5205.
38. Lovato N, Lack L, Wright H, Cant M, Humphreys J. Working memory performance of older adults with insomnia. *J Sleep Res.* 2013;22(3):251–257.
39. Haimov I, Hanuka E, Horowitz Y. Chronic insomnia and cognitive functioning among older adults. *Behav Sleep Med.* 2008;6(1):32–54.
40. Caplette-Gingras A, Savard J, Savard MH, Ivers H. Is insomnia associated with cognitive impairments in breast cancer patients? *Behav Sleep Med.* 2013;11(4):239–257.
41. Deckersbach T, Moshier SJ, Tuschen-Caffier B, Otto MW. Memory dysfunction in panic disorder: an investigation of the role of chronic benzodiazepine use. *Depress Anxiety.* 2011;28(11):999–1007.
42. Saloner R, Grelotti DJ, Tyree G, et al. Benzodiazepine use is associated with an increased risk of neurocognitive impairment in people living with HIV. *J Acquir Immune Defic Syndr.* 2019;82(5):475–482.
43. Rapeli P, Fabritius C, Kalska H, Alho H. Memory function in opioid-dependent patients treated with methadone or buprenorphine along with benzodiazepine: longitudinal change in comparison to healthy individuals. *Subst Abuse Treat Prev Policy.* 2009;4:6.
44. Fond G, Berna F, Boyer L, et al. Benzodiazepine long-term administration is associated with impaired attention/working memory in schizophrenia: results from the national multicentre FACE-SZ data set. *Eur Arch Psychiatry Clin Neurosci.* 2018;268(1):17–26.
45. Verwey B, Eling P, Wientjes H, Zitman FG. Memory impairment in those who attempted suicide by benzodiazepine overdose. *J Clin Psychiatry.* 2000;61(6):456–459; quiz 460.
46. Verwey B, Muntendam A, Ensing K, et al. Clinically relevant anterograde amnesia and its relationship with blood levels of benzodiazepines in suicide attempters who took an overdose. *Prog Neuropsychopharmacol Biol Psychiatry.* 2005;29(1):47–53.
47. Boeuf-Cazou O, Bongue B, Ansiau D, Marquié JC, Lapeyre-Mestre M. Impact of long-term benzodiazepine use on cognitive functioning in young adults: the VISAT cohort. *Eur J Clin Pharmacol.* 2011;67(10):1045–1052.
48. Barker MJ, Greenwood KM, Jackson M, Crowe SF. Cognitive effects of long-term benzodiazepine use: a meta-analysis. *CNS Drugs.* 2004;18(1):37–48.

13 Naturally Occurring and Exogenous Benzodiazepines in Epilepsy: An Update

Felix-Martin Werner
Höhere Berufsfachschule für Altenpflege
und Ergotherapie der Euro Akademie Pößneck
University of Salamanca
Instituto de Neurosciencias de Castilla y León (INCYL)

Rafael Coveñas
University of Salamanca
Instituto de Neurosciencias de Castilla y
León (INCYL) GIR USAL group: BMD

CONTENTS

INTRODUCTION

Benzodiazepines play an important role in different medical fields because they promote sedative and sleep-inducing, anxiolytic, muscle-relaxing, and antiepileptic effects. Benzodiazepine receptors are allosteric modulatory binding sites of the gamma-aminobutyric acid $(GABA)_A$ receptor complex, which is located in the plasma membrane of various neurons (Sieghart, 1990). GABA released by GABAergic neurons, $GABA_A$ agonists, and benzodiazepines exerts an inhibitory function on postsynaptic neurons, including dopaminergic and noradrenergic neurons, via the activation of $GABA_A$ receptors distributed in different brain regions (e.g., brainstem, hippocampus, prefrontal cortex, extrapyramidal system). Some benzodiazepines (e.g., diazepam, lorazepam) (Figure 13.1) not only have a beneficial effect during the long-term treatment of generalized epilepsy, but also interrupt the acute grand-mal seizure.

Here, an overview of the endogenous and exogenous benzodiazepines exerting an antiepileptic effect will be given, including their pharmacokinetics, pharmacodynamics, and specific actions. Oleamides, non-peptidic endozepines, and the diazepam-binding inhibitor (DBI) are endogenous benzodiazepines. In order to describe the function of endozepines, their interactions with the benzodiazepine-binding site at the $GABA_A$ receptor and their roles in influencing epileptic seizures will be described. DBI displaces synthetic benzodiazepines from the benzodiazepine-binding site at the $GABA_A$ receptor, but it can as well exert a positive allosteric modulation of this receptor (Farzampour et al., 2015). Thus, the different forms of application, the indications, and the appropriate dosages will be indicated as well as the adverse effects of benzodiazepines (Heller et al., 2018; Rogawski and Heller, 2019). It is important to note that, in a status epilepticus, benzodiazepines lose their antiepileptic effects when the acute seizure lasts for more than 20 min. In these cases, benzodiazepines should be combined with glutamatergic antiepileptic drugs as a consequence of the receptor trafficking (Werner and Coveñas, 2011). This will be also mentioned in the corresponding section. In generalized epilepsy, a neurotransmitter imbalance between hypoactive $GABA_A$ GABAergic neurons and hyperactive N-methyl-D-aspartate (NMDA) glutamatergic neurons has been

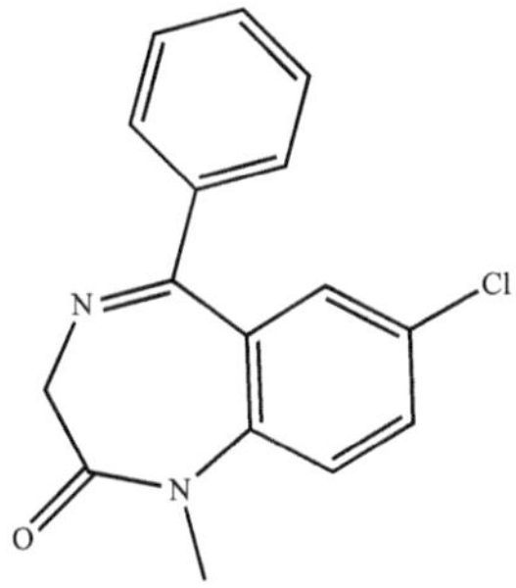

DIAZEPAM

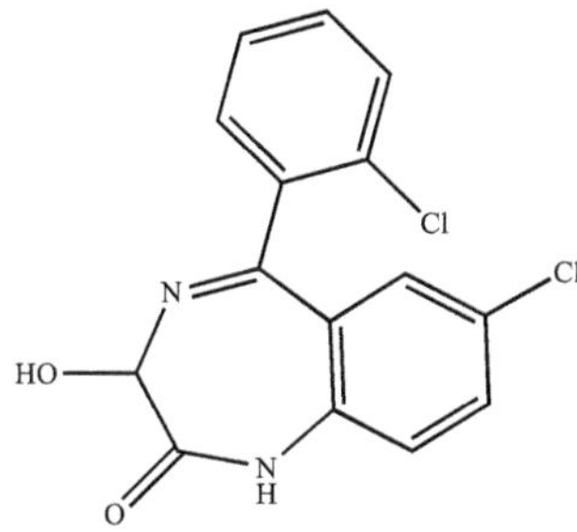

LORAZEPAM

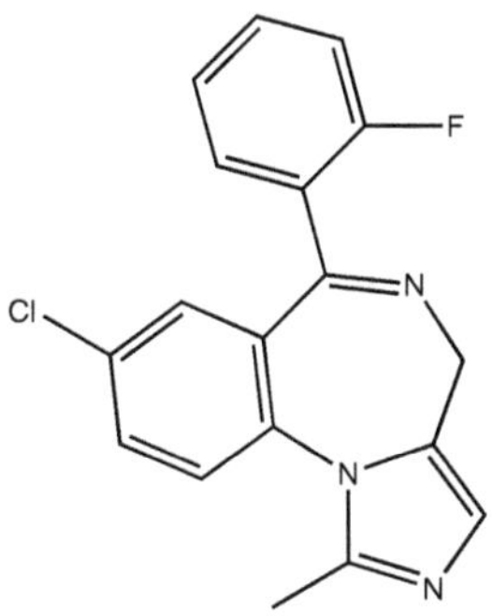

MIDAZOLAM

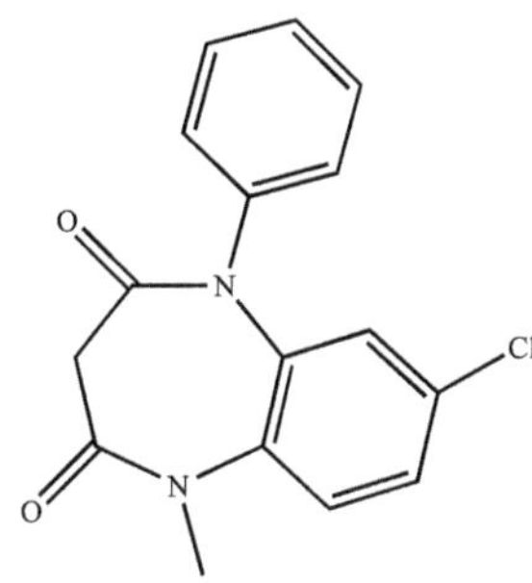

CLOBAZAM

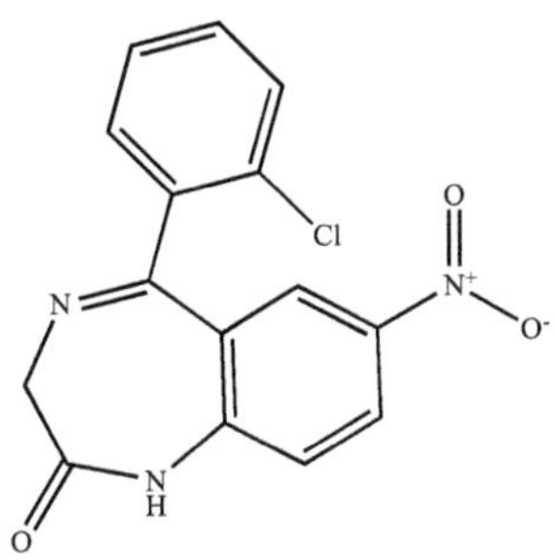

CLONAZEPAM

FIGURE 13.1 Chemical structures of antiepileptic benzodiazepines: diazepam, clobazam, clonazepam, lorazepam, and midazolam.

reported in the hippocampus (Werner and Coveñas, 2017a). Here, the corresponding neurotransmitter alterations of dopaminergic, serotonergic, GABAergic, and glutamatergic neurons will be described, and in generalized epilepsy, the involved neural networks located in the hippocampus, thalamus, and cerebral cortex will be updated.

HISTORY OF EXOGENOUS AND NATURALLY OCCURRING BENZODIAZEPINES

The first benzodiazepine, chlordiazepoxide, was synthesized by the chemist Leo Sternbach in 1955. After its synthesis, many pharmaceutical companies modified the molecular structure of chlordiazepoxide, and then, a wide range of benzodiazepines was developed. Diazepam (Figure 13.1), which is being used still today as an emergency antiepileptic drug, was developed in 1960. Benzodiazepines do not cause respiratory depression and have the advantage to be less toxic than barbiturates (Wick, 2013). Only in the mid- to late 1970s, the agonistic effect of benzodiazepines on GABA receptors was described. During this period, benzodiazepines were among the most frequently prescribed drugs. However, two adverse effects limited the wide use of this group of drugs. On the one hand, they can cause dependence, which remains difficult to be cured, and on the other hand, they exert a paradoxical reaction in elderly patients because, instead of having a sedative effect, they make patients more active (Wick, 2013). However, benzodiazepines have a therapeutic effect in some forms of epilepsy and are used as secure emergency drugs for the treatment of seizures. In this case, the (parental vs nonparental) route of administration and the pharmacokinetic profile of benzodiazepines must be well established (Rogawski and Heller, 2019).

Naturally occurring benzodiazepines were discovered in 1983. They are supposed to exert both a negative and a positive allosteric modulation of the benzodiazepine-binding site at the $GABA_A$ receptor. Endozepines such as DBI and its cleavage products triakontatetraneuropeptide (TTN), octadecaneuropeptide (ODN), and octaneuropeptide (OD) have been reported. These endozepines can displace administered benzodiazepines from the benzodiazepine-binding site at the $GABA_A$ receptor. DBI can exert a negative allosteric modulation and, at high doses, a positive allosteric modulation of this receptor (Farzampour et al., 2015).

MECHANISMS OF ACTION OF BENZODIAZEPINES

Benzodiazepines

Benzodiazepines are positive allosteric modulators of $GABA_A$ receptors, and although their binding to the receptor is classically antagonized by flumazenil, recent findings have highlighted the occurrence of flumazenil-insensitive effects (Lian et al., 2020). In mice, it has been investigated the effect of progesterone in an immobilization-stress situation and the involvement of $GABA_A$ and mitochondrial DBI receptors in this effect. The immobilization of mice for 2h led to antinociception, anxiety state, and a decrease in adrenal ascorbic acid level. These stress-induced symptoms were abolished by pretreatment with a high dose of progesterone, a precursor of neurosteroids. The antistress effect of progesterone was antagonized by picrotoxin and bicuculline, two $GABA_A$ receptor antagonists, but not by flumazenil, a specific benzodiazepine receptor antagonist (Reddy and Kulkarni, 1996). Moreover, 4'-diazepam promoted antistress effects, which were insensitive to flumazenil but were blocked by PK-11,195, a selective partial agonist of the mitochondrial DBI receptor, and by bicuculline, a potent $GABA_A$ receptor antagonist (Reddy and Kulkarni, 1996). The antistress effect of progesterone may be due to the in vivo production of neuroactive steroids that may act on a specific allosteric site of $GABA_A$ receptors distinct from the benzodiazepine-binding site (Reddy and Kulkarni, 1996).

In mouse brain slices, the action of a photo-switchable form of diazepam, namely, photocaged version of diazepam, was investigated on basolateral amygdala pyramidal cells. Caged diazepam is a photoreleasable $GABA_A$ receptor-positive allosteric modulator, which shows different properties from caged GABA or other photo-switchable ligands. Caged diazepam enhanced the inhibitory signaling by prolonging the decay time of postsynaptic GABAergic currents through photoactivation (Sansalone et al., 2019).

Hyperthermia-induced febrile seizures are the most common epileptic seizures in childhood, and prolonged epileptic seizures can later lead to epilepsy. An important question to solve is whether benzodiazepines or barbiturates, which activate GABAergic neurotransmission, can depolarize immature neurons with a relatively high level of intracellular chloride. The antiepileptic effect of benzodiazepines and barbiturates has been examined in postnatal P11 and P14 mice. In P11, administration of a low dose of diazepam decreased the occurrence of clonic seizures. However, in P11 and P14, high doses of diazepam and pentobarbital exacerbated behavioral and electrophysiological phenotypes in the induction phase of febrile seizures. The deterioration of the seizures was prevented by the diuretic bumetanide, which inhibited the chloride influx (Kasahara et al., 2019).

In a retrospective study, patients suffering from a status epilepticus were examined in order to find out the adequate dose of benzodiazepines to be administered. This is important because administering the adequate dose, the status epilepticus, i.e., an epileptic seizure that lasts more than 5 min or when several seizures occur with recovering consciousness, should not progress to refractory status epilepticus or to non-convulsive status epilepticus. It was reported that the adequate dose of lorazepam (Figure 13.1) was between 2 and 5 mg, since patients with status epilepticus who did not receive this adequate dosage in many cases progressed to refractory status epilepticus or to non-convulsive status epilepticus with coma (Rao et al., 2018).

Endozepines

After the use of benzodiazepines in clinical medicine since the 1960s and the discovery of the benzodiazepine site at the $GABA_A$ receptor, a question arises: whether endogenous ligands on this binding site might exist. This assumption was proved after the discovery of oleamides, non-peptidic endozepines, and the protein DBI. Several studies showed that endozepines/naturally occurring benzodiazepines exerted a negative allosteric modulation of the benzodiazepine site at the $GABA_A$ receptor, but recent studies revealed that DBI and other endozepines exerted as well a positive allosteric modulation of this receptor (Farzampour et al., 2015).

Mechanisms of Action of Endozepines

Because benzodiazepines, like diazepam, are positive allosteric modulators of the $GABA_A$ receptor by enhancing the chloride current, a question arises: which are the pharmacodynamic effects exerted by endozepines? Ligands such as beta-carbolines inhibit the action of GABA and are considered inverse agonists or negative allosteric modulators of the benzodiazepine-binding site. Ligands such as flumazenil inhibit the effect of positive and negative allosteric modulators, and they have no intrinsic effect on GABA. Consequently, these ligands act as benzodiazepine receptor antagonists (Farzampour et al., 2015); however, flumazenil has not been proven to exert antagonistic effects on the benzodiazepine receptor, and hence, it was hypothesized that this antagonistic effect was performed via endogenous ligands (Ramerstdorfer et al., 2010). Flumazenil suppresses inhibitory postsynaptic potentials in different neuronal circuits, and in *in vitro* studies using flumazenil, it has been shown that endogenous ligands might also have a positive allosteric modulation (Farzampour et al., 2015). Flumazenil can aggravate epileptic seizures, while it has a negative allosteric modulation at low doses and a positive allosteric modulation of the benzodiazepine site at the $GABA_A$ receptor at higher concentrations (Rye et al., 2012).

Diazepam-Binding Inhibitor

DBI is the most widely studied endozepine. It was found in neurons and astrocytes of the mammalian brain and displaces synthetic benzodiazepines from the benzodiazepine-binding site at the $GABA_A$ receptor (Alho et al., 1991). DBI was found to be identical to the acyl-CoA-binding protein, which plays a role in the fatty acid metabolism (Knudsen, 1991). DBI synthesis and excretion is mostly found in astrocytes (Loomis et al., 2010). Ten different transcript variants of DBI have been found, and their functions (e.g., fatty acid metabolism, variation in the function of the $GABA_A$ receptor) have not yet been clearly determined (Nitz et al., 2011).

A question arises, whether DBI exerts a positive or negative allosteric modulation at the $GABA_A$ receptor. The exogenous application of DBI reduced the presynaptic inhibitory action at the $GABA_A$ receptor. However, the excretion of DBI is necessary for a $GABA_A$ receptor positive allosteric modulation by endozepines. Consequently, DBI can perform a positive and negative modulation at the $GABA_A$ receptor (Farzampour et al., 2015).

Diazepam-Binding Inhibitor and Its Cleavage Products: Triakontatetraneuropeptide (TTN), Octadecaneuropeptide (ODN), and Octapeptide (OD)

DBI has several cleavage products, including TTN, ODN, and OD, which have been identified in the rat and rhesus monkey central nervous system (CNS) and have the ability to displace benzodiazepines from their binding site at the $GABA_A$ receptor and also modulate the allosteric inhibition of the $GABA_A$ receptor (Alho et al., 1991). OD suppresses pentylenetetrazol (PTZ)-induced seizures, but at high doses promotes convulsant effects (Ferrero et al., 1986). Intracerebroventricular injections of DBI, TTN, and OD induced anxiogenic activity in rats through the benzodiazepine-binding site at the $GABA_A$ receptor, and this effect could be antagonized by flumazenil (Slobodyansky et al., 1989).

In generalized epilepsy, endozepines have a regulating function at the hippocampal neural circuits although the exact mechanism of action remains unclear.

Endozepines such as DBI and TTN can as well bind to a peripheral benzodiazepine receptor, which is a cholesterol transporter located in the outer mitochondrial membrane in non-neuronal tissues and neurons (Slobodyansky et al., 1989). The binding of endozepines to the peripheral benzodiazepine receptor stimulates cholesterol transport into mitochondria and influences neurosteroid synthesis. Neurosteroids bind to an allosteric site of the $GABA_A$ receptor, which is different from the benzodiazepine-binding site, and enhance synaptic and extrasynaptic functions of the $GABA_A$ receptor (Porcello et al., 2003).

Oleamide

Oleamide (cis-9,10-octadecenoamide) is an endogenous neuroactive substance which accumulates in sleep deprivation and induces sleep in animals. In cell cultures, oleamide is a modulator of presynaptic and voltage-gated Na^+ channels and of postsynaptic $GABA_A$ receptors. Oleamide exerts effects similar to barbiturates, which display an anticonvulsant effect (Verdon et al., 2000).

Astrocytes Potentiate GABAergic Presynaptic Inhibition in the Thalamic Reticular Nucleus Via Endozepine Signaling

It has been described that DBI displays a benzodiazepine-like effect on the GABAergic presynaptic inhibition in the thalamic reticular nucleus. In this nucleus, DBI is localized in both neurons and astrocytes. After the administration of gliotoxin fluorocitrate (FC) and using patch-clamp recordings, the involvement of astrocytes in enhancing the GABAergic presynaptic inhibition was examined (Christian and Huguenard, 2013). Treatment with FC reduced the presynaptic inhibition exerted by GABAergic neurons through $GABA_A$ receptors in control mice, and this indicates that astrocytes enhance the positive allosteric modulation of the $GABA_A$ receptor through endogenous endozepines. This effect was blocked by a deletion of the DBI gene and by a point mutation that inhibits benzodiazepine binding to the alpha3 subunit of the $GABA_A$ receptor, which is predominant in the thalamic reticular nucleus. The ventrobasal nucleus of the thalamus is not involved in this effect, because it does not contain endogenous benzodiazepines (Christian and Huguenard, 2013).

MULTI-NEUROTRANSMITTER SYSTEM: IMPORTANCE OF GABA, GLUTAMATE, DOPAMINE, AND SEROTONIN IN GENERALIZED EPILEPSY

In generalized epilepsy, the involvement of a multi-neurotransmitter system has been reported in the hippocampus (Werner and Coveñas, 2015). It has been suggested that the hypoactivity of the GABAergic neurotransmission (GABAergic neurons exert a presynaptic inhibitory function upon dopaminergic or noradrenergic neurons via $GABA_A$ receptors) and the hyperactivity of the glutamatergic neurons (which perform an excitotoxic effect, namely, a postsynaptic excitatory function and partly a presynaptic inhibitory function) enhance ictogenesis (Werner and Coveñas, 2015). $GABA_A$ antagonists and glutamate can increase the electrographic seizure activity, whereas $GABA_A$ agonists or benzodiazepines and NMDA receptor antagonists exert an antiepileptic effect (Butuzova and Kitchigina, 2008). Endozepines (e.g., DBI) can also induce or suppress epileptic seizures, because DBI displaces benzodiazepines from the benzodiazepine-binding site at the $GABA_A$ receptor. At high doses, DBI can exert a positive allosteric modulation of the $GABA_A$ receptor. Through this mechanism, the GABAergic presynaptic inhibition of dopaminergic neurons is reduced, and this can promote proconvulsant effects (Farzampour et al., 2015). In generalized epilepsy, dopaminergic neurons influence ictogenesis via D_2 receptors because dopamine hyperactivity can induce epileptic seizures. In limbic seizures, a selective dopamine reuptake inhibitor (GBR-12,909) exerted antiepileptic effects through D_2 receptors (Clinckers et al., 2004). It is important to remark that antipsychotic drugs displaying a D_2 antagonistic effect can enhance proneness to epileptic seizures. These drugs block D_2 receptors and activate the presynaptic synthesis of dopamine. Among the second-generation antipsychotic drugs, clozapine, which exerts a D_3, D_4, and 5-HT_{2A} antagonistic effect, often induces epileptic seizures as an adverse effect (Agrawal and Mula, 2019). According to the National Institute of Health and Care Excellence, the second-generation antipsychotic drugs risperidone, olanzapine, and quetiapine have the most favorable profile of adverse effects (Agrawal and Mula, 2019). Serotonin, a major postsynaptic excitatory neurotransmitter, plays as well an important role in ictogenesis. Serotonin hypoactivity exerts a proconvulsant effect, while selective serotonin reuptake inhibitors (e.g., citalopram) perform a slight antiepileptic function through the activation of 5-HT_{1A} receptors (Thomé-Souza et al., 2007). The neuroactive substance adenosine is as well involved in ictogenesis because, via A_{2A} receptors, exerts a hyperactive effect (Werner and Coveñas, 2017b). In the future, an important question must be answered, whether A_{2A} antagonists exert or not an antiepileptic effect.

GENERALIZED EPILEPSY: NEURAL NETWORKS IN HIPPOCAMPUS, THALAMUS, AND CEREBRAL CORTEX

Adapted from the previously published works (Werner and Coveñas, 2011, 2015, 2017a, b, 2019), updated neural networks located in the hippocampus, thalamus, and cerebral cortex and involved in generalized epilepsy are shown in Figure 13.2. In the hippocampus, D_2 dopaminergic neurons perform a postsynaptic excitatory potential with a high activity upon glutamatergic neurons, which presynaptically inhibit 5-HT_{2A} serotonergic neurons via NMDA receptors. Glutamatergic neurons exert as well an excitotoxic effect via NMDA, alpha-amino-3-hydroxy-5-methyl-4-isoxazole-propionic acid (AMPA), and kainate (KA) receptors upon D_2 dopaminergic neurons. 5-HT_{2A} serotonergic neurons transmit a weak postsynaptic potential upon GABAergic neurons, which apply a weak presynaptic inhibitory potential via $GABA_A$ receptors upon D_2 dopaminergic neurons. Thus, the withdrawal of the GABAergic presynaptic inhibition (mediated by $GABA_A$ receptors) on D_2 dopaminergic neurons can induce epileptic seizures. As indicated above, endozepines (e.g., DBI) can induce epileptic seizures because the GABAergic presynaptic inhibition exerted on D_2 dopaminergic neurons is reduced (Farzampour et al., 2015). Hippocampal GABAergic neurons inhibit glutamatergic neurons located in the thalamus and the latter neurons activate, via NMDA receptors, cortical glutamatergic neurons, which can activate hippocampal D_2 dopaminergic neurons (Werner and Coveñas, 2015, 2019). 5-HT_{2A} serotonergic neurons (with a weak activity) also transmit a weak postsynaptic potential to other hippocampal GABAergic neurons, which weakly presynaptically inhibit A_{2A} adenosine

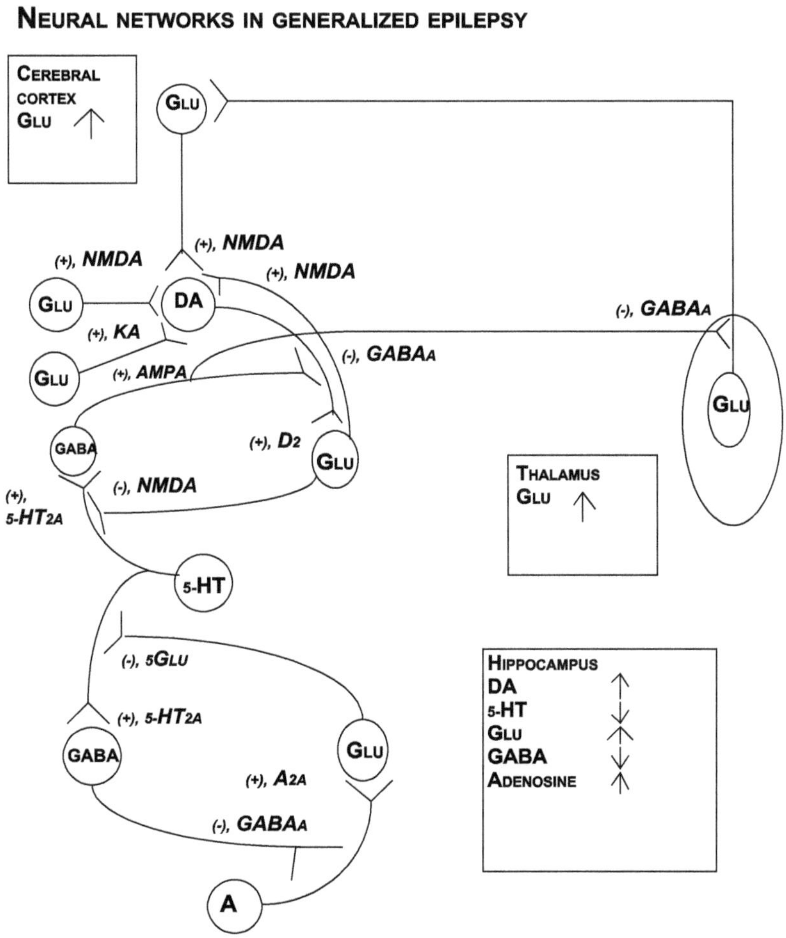

FIGURE 13.2 Generalized epilepsy: neural networks in the hippocampus, thalamus, and cerebral cortex (updated from Werner and Coveñas, 2011, 2015, 2017a, b, 2019). 5-HT: serotonin; A: adenosine; DA: dopamine; GABA: gamma-aminobutyric acid; Glu: glutamate. The following receptor subtypes are depicted: A_{2A}: adenosine type 2A receptor subtype; AMPA: ionotropic glutamatergic AMPA receptor subtype; $GABA_A$: GABA receptor subtype A; 5Glu: metabotropic glutamatergic receptor subtype 5; 5-HT_{2A}: serotonergic receptor subtype 2A; D_2: type 2 dopamine receptor; KA: kainate ionotropic glutamatergic receptor; NMDA: *N*-methyl-D-aspartate ionotropic glutamatergic receptor. A plus mark indicates a postsynaptic excitatory impulse; a minus mark indicates a presynaptic inhibitory impulse. ↑: increase; ↓: decrease.

neurons via $GABA_A$ receptors. The latter neurons (with a high activity) apply a strong postsynaptic excitatory potential to glutamatergic neurons, which exert a strong presynaptic inhibitory effect on 5-HT_{2A} serotonergic neurons via subtype 5 of metabotropic glutamatergic receptors (Werner and Coveñas, 2015, 2017a, 2019). According to the neural networks described, benzodiazepines exert a positive allosteric effect on $GABA_A$ receptors. Through this mechanism of action, the GABAergic presynaptic inhibitory action reduces the postsynaptic excitatory potential of D_2 dopaminergic neurons; the latter neurons weakly activate glutamatergic neurons. Consequently, the glutamatergic presynaptic inhibitory action of NMDA glutamatergic neurons on 5-HT_{2A} serotonergic neurons is reduced. Due to this action, the hypoactivity of 5-HT_{2A} serotonergic neurons is decreased. Thus, the pharmacological effect induced on different neuronal populations enhances the antiepileptic effect.

RECEPTOR TRAFFICKING IN STATUS EPILEPTICUS: APPROPRIATE APPLICATION OF BENZODIAZEPINES AND NMDA RECEPTOR ANTAGONISTS

When single epileptic seizures become a status epilepticus, benzodiazepines administered as a monotherapy often lose their effectiveness. This might be due to $GABA_A$ and NMDA glutamate receptor trafficking. During the status epilepticus, $GABA_A$ receptors are internalized; i.e., they move from extracellular to intracellular sites, whereas NMDA glutamate receptors move from intracellular to extracellular sites (Niquet et al., 2016). The following question arises: should it be advisable to combine benzodiazepines with an NMDA receptor antagonist when epileptic seizures become a status epilepticus? To examine this, a status epilepticus was induced in an animal model of epilepsy after the administration of high doses of lithium and pilocarpine and as well of subcutaneous soman. Monotherapy with the $GABA_A$ receptor positive allosteric modulator midazolam led to a refractory status epilepticus. However, the simultaneous administration of midazolam (Figure 13.1) with the NMDA receptor antagonist dizocilpine or ketamine successfully treated the refractory status epilepticus, independently of the dosage of the epileptogenic drug administered. The subsequential administration of both drugs (midazolam and an NMDA receptor antagonist) was less successful than the simultaneous administration. The authors recommend a three-dimensional therapy with midazolam, ketamine, and valproate (Niquet et al., 2019). The influence of endozepines (e.g., DBI) and its cleavage products on the course of a longer-lasting epileptic seizure has not yet been examined (Farzampour et al., 2015).

To determine the appropriate administration of drugs to treat the status epilepticus, this state was induced in adult Sprague-Dawley rats after the administration of high doses of lithium and pilocarpine. The dual therapy with midazolam and ketamine was more effective than the monotherapy with midazolam or ketamine, the combination of midazolam with valproate, or that of ketamine with valproate. Because during the course of a convulsive status epilepticus, $GABA_A$ receptors are internalized and NMDA glutamate receptors are externalized, a combination therapy using a positive allosteric modulator of the $GABA_A$ receptor and an NMDA receptor antagonist is advisable (Niquet et al., 2016).

In a review based on animal experiments and case reports about convulsive status epilepticus, the authors recommended to apply non-benzodiazepine antiepileptic drugs, i.e., NMDA receptor antagonists, such as ketamine, for the treatment of seizures lasting longer than 5 min to prevent complications such as hyperthermia, hypoglycemia, hypoxia, and cardiorespiratory instability (Sánchez Fernández et al., 2019).

PHARMACOKINETIC PROFILES OF BENZODIAZEPINES

In this section, the pharmacokinetic profiles of endozepines and of naturally occurring benzodiazepines and synthetic benzodiazepines showing an antiepileptic action (namely, diazepam, lorazepam, midazolam, clobazam, and clonazepam) (Figure 13.1) will be mentioned. Hints for an appropriate administration of these benzodiazepines during the management of epileptic seizures will be also given.

Diazepam

In a multicenter, double-blind, and randomized study, children (aged between 3 months and 18 years) who suffered from status epilepticus with tonic–clonic seizures were selected (Table 13.1). They received an intravenous injection of diazepam (0.2 mg/kg up to a maximum of 8 mg; over a minute), and the results obtained were compared with those found when lorazepam (Figure 13.1) was intravenously administered (Ku et al., 2018). In the case that epileptic seizures continued, a second dose of diazepam (0.1 mg/kg) was administered and, in case of seizure persistence, an additional dose of phenytoin and/or phenobarbital was administered. In this study, 87 children were enrolled, and 24 of them received a second dose. A therapeutic dose of 200–600 mg/ml was reached after one or two injections. At the dose indicated above, diazepam performed an antiepileptic effect. Liver parameters were within normal ranges. Diazepam suffers a liver transformation, and its major metabolite, desmethyldiazepam, only has a weak intrinsic activity (Ku et al., 2018). This study confirmed that the therapeutic dose of 200–600 mg/ml was reached after 10 min of receiving the first injection of diazepam (Ku et al., 2018).

In a phase I, open-label, randomized study with 48 healthy subjects, the nasal administration of diazepam was compared to the oral or rectal gel administration (Hogan et al., 2020). After the nasal application, diazepam was absorbed rapidly. The $f_{\text{max.}}$ (time to reach the maximum plasma concentration) of intranasal diazepam was similar to that of rectal diazepam, but both routes of administration were slower than oral diazepam. The lowest variability in diazepam plasma concentration was observed after the oral administration, followed by the intranasal diazepam, whereas the greatest variability was found after the rectal application. During the study, some mild and moderate treatment-emergent adverse effects were described; however, there were no severe reactions. The most common adverse effect was somnolence (this was observed in 56.5% of patients who received the intranasal diazepam). The latter application showed a good pharmacokinetic profile and exerted a therapeutic effect in controlling epileptic seizures (Hogan et al., 2020).

TABLE 13.1
Benzodiazepines Showing Antiepileptic Effects (Diazepam, Clobazam, Clonazepam, Lorazepam, Midazolam): Indications, Routes of Administration, and Doses

Benzodiazepines	Indications	Route of Administration	Doses	References
Diazepam	Epileptic seizure	Intranasal	15–20 mg (single dose)	Hogan et al. (2020)
	Status epilepticus	Intravenous	0.15 mg/kg	Mackway-Jones (2001)
	Status epilepticus (children)	Intravenous	0.2–8 mg/kg	Ku et al. (2018)
	Epileptic seizure	Oral	2–8 mg/day; maximum: 30 mg/day	Calcaterra and Barrow (2014)
Clobazam	Epileptic seizure	Oral	10–40 mg/day	Huddart et al. (2018)
	Lennox–Gestaut syndrome	Oral	10–40 mg/day	Huddart et al. (2018)
Clonazepam	Epileptic seizure or status epilepticus	Buccal	1 mg	Schols-Hendriks et al. (1995)
	Epileptic seizure or status epilepticus	Intranasal	1 mg	Schols-Hendriks et al. (1995)
	Epileptic seizure or status epilepticus	Intravenous	0.5 mg	Schols-Hendriks et al. (1995)
Lorazepam	Epileptic seizure or status epilepticus	Intravenous	0.1–5 mg	Butler and Lewis (2001) and Rao et al. (2018)
	Epileptic seizure or status epilepticus	Oral	0.1–5 mg	Butler and Lewis (2001) and Rao et al. (2018)
Midazolam	Epileptic seizure or status epilepticus	Intranasal	5 mg	Knoester et al. (2002)
	Epileptic seizure or status epilepticus	Intravenous	5 mg	Knoester et al. (2002)
	Epileptic seizure or status epilepticus	Oral	15 mg	Wang et al. (2016)

Lorazepam

It is important to know whether lorazepam or diazepam could be administered when an adult suffers epileptic seizures and when an intravenous access is secured. The duration of the action of diazepam is about 20–30 min, and it shows a widespread distribution by and outside the nervous system, whereas lorazepam has an elimination half-life of 13 h. At the beginning of a status epilepticus, lorazepam (0.1 mg/kg) or diazepam (0.15 mg/kg) is usually administered. This status shows a frequency of 15–30 of 100.000 seizures a year (Butler and Lewis, 2001). Lorazepam (intravenous injection) is a safe antiepileptic drug when the epileptic seizure lasts longer than 5 min (Butler and Lewis, 2001).

Midazolam

The pharmacokinetic and pharmacodynamic of midazolam (Figure 13.1), administered intravenously or as a concentrated intranasal spray, have been examined in healthy subjects. Midazolam can be administered to epileptic patients suffering from a status epilepticus. Subjects received an intravenous or intranasal administration of 5 mg midazolam. Plasma concentrations of midazolam and its major metabolite, 1-hydroxy-midazolam, were determined; in addition, β-band EEG activity was recorded and related to the midazolam plasma concentration (Knoester et al., 2002). All subjects receiving the intranasal administration of midazolam felt irritated, but tolerated well this application. The peak plasma concentration of 17 ng/ml midazolam was reached after 14 min in the average, and the mean bioavailability of intranasally administered midazolam was 0.83. The mean systematic clearance was 16.1 ml/min, and the terminal half-life was 79 min. After an intranasal administration, the plasma concentration of 1-hydroxy-midazolam was not higher than that found after an intravenous application. From this phase I study, it can be concluded that the administration of concentrated intranasal midazolam, like an intravenous injection, can be applied to treat the status epilepticus (Knoester et al., 2002).

Clobazam

Clobazam (Figure 13.1) is a benzodiazepine approved by the Food and Drug Administration for the treatment of the Lennox–Gestaut syndrome in children older than 2 years. Besides, it can be used for the treatment of refractory epilepsy. Clobazam binds to postsynaptic $GABA_A$ receptors and promotes hyperpolarization. The inhibitory potential is thereby enhanced, and the threshold for epileptic seizures is also increased. The major metabolite of clobazam, norclobazam, also exerts an antiepileptic effect. Clobazam should not be administered with lamotrigine, because this combination therapy increases the risk for allergic reactions such as the Steven–Johnson syndrome (Huddart et al., 2018). Clobazam has a half-life of 36–42 h, and that of its major metabolite, norclobazam, 71–82 h. Clobazam reaches the steady-state concentration after 1 week and norclobazam after 3 weeks. High concentrations of clobazam were found in patients treated with a regular administration of the drug.

Consequently, norclobazam serum concentration is important for the therapeutic effect of clobazam. T_{max} of clobazam is reached after 0.5–3 h, and that of norclobazam after 3–56 h. Clobazam has an extensive liver metabolism of about 70% and, in patients with a liver impairment, it is recommended to reduce its dose. Clobazam is excreted through the urine and feces (Huddart et al., 2018).

Clonazepam

The benzodiazepine clonazepam (Figure 13.1) can be used for the treatment of the status epilepticus or serial seizures (Table 13.1). In a phase I study, buccal, intranasal, and intravenous administration of clonazepam was examined in seven male healthy volunteers aged between 26 and 58 years (Schols-Hendriks et al., 1995). Each participant received 1.0 mg of clonazepam intranasally/buccally and 0.5 mg intravenously. A C_{max} of 6.3 ng/ml of clonazepam was reached 17.5 min after the intranasal administration was performed. A second peak of 4.6 ng/ml caused by oral absorption was reached after 1.7 h. A C_{max} of 6.0 ng/ml was reached 50 min after the buccal administration, and a second peak of 6.5 ng/ml was found after 3.0 h. Two minutes after the intravenous administration, a serum concentration of clonazepam of 27 ng/ml was measured. The intranasal administration of clonazepam is an alternative route for the buccal administration; however, to treat a status epilepticus, the intranasal administration of clonazepam cannot be replaced by its intravenous injection (Schols-Hendriks et al., 1995).

INDICATIONS OF BENZODIAZEPINES IN EPILEPSY

Among the benzodiazepines, diazepam, lorazepam, midazolam, clobazam, and clonazepam can be administered to treat generalized epilepsies, acute epileptic seizures, or status epilepticus (Table 13.1). Diazepam can be used to treat acute epileptic seizures and status epilepticus. If epileptic seizures last over 5 min, and if it can become a status epilepticus, an additional administration of NMDA receptor antagonists such as ketamine is recommended as a consequence of the $GABA_A$ and NMDA receptor trafficking (Niquet et al., 2019). Lorazepam can be applied to treat epileptic seizures and a threatening status epilepticus. It can be administered by different routes, for example, by an intravenous injection or by an oral administration (Butler and Lewis, 2001). Midazolam, which can be administered as an intravenous injection or as a concentrated intranasal application, can be used to treat the status epilepticus. Both intravenous and intranasal routes of administration enable a therapeutic plasma concentration of the antiepileptic drug (Knoester et al., 2002). Clobazam, showing a half-life of 36–42 h, has been approved for the treatment of the Lennox–Gestaut syndrome in children older over 2 years. Moreover, clobazam can be chosen to treat refractory epilepsies (Huddart et al., 2018). Clonazepam is indicated for the treatment of the status epilepticus and serial seizures. Buccal, intranasal, and intravenous administrations are appropriate routes (Schols-Hendriks et al., 1995).

ROUTES OF ADMINISTRATION AND APPROPRIATE DOSAGES

In this section, to treat generalized epilepsies and status epilepticus, the routes of administration and appropriate dosages of the above-mentioned five benzodiazepines will be described (Table 13.1). In children, to treat the status epilepticus, diazepam (intravenous injection) is dosed from 0.2 mg/kg to a maximum of 8 mg/kg. To treat epileptic seizures, between 2 and 8 mg/day of diazepam (with a maximum of 30 mg/day) can be given by the oral route (Calcaterra and Barrow, 2014). As an intranasal application, diazepam is dosed from 15 to 20 mg in a single dose (Hogan et al., 2020). In a post hoc study and to treat the status epilepticus, the effects of the intravenous injection of diazepam (0.15 mg/kg) or lorazepam (0.1 mg/kg) were compared: the findings indicated that the therapeutic effect of lorazepam was safer and more tolerable than that of diazepam (Mackway-Jones, 2001). Midazolam can be administered to treat epileptic seizures and status epilepticus: midazolam can be given orally (15 mg) or by the intravenous or intranasal route (5 mg) (Knoester et al., 2002; Wang et al., 2016). Clobazam can be administered to treat the Lennox–Gestaut syndrome or epileptic seizures: the oral dose is 10–40 mg/day with an average dosage of 30 mg/day (Huddart et al., 2018). Clonazepam can be the choice to treat status epilepticus or serial seizures. In this indication, the appropriate route of administration can be buccal or intranasal (1 mg) or intravenous (0.5 mg) (Schols-Hendriks et al., 1995).

ADVERSE EFFECTS OF BENZODIAZEPINES

In general, adverse effects of benzodiazepines are somnolence, dependence, amnesia, and a reduced reaction time. Moreover, the observed adverse effects of benzodiazepines are tiredness, hypotension with vertigo, ataxia, dysarthria, and a reduced sexual function. Known risk factors for adverse effects of benzodiazepines are lung disease, substance abuse, and vulnerability to fracture (Kroll et al., 2016). Less frequent adverse effects are increased weight and appetite, irritations of the bowels, edema, allergic skin reactions, muscle weakness, and double images (Tölle and Windgassen, 2003). In a chronic misuse of benzodiazepines, patients can develop deliriant psychoses (Tölle and Windgassen, 2003). In elderly patients, benzodiazepines can cause a paradoxical reaction characterized by an activating action instead of a soothing and antiepileptic effect (Wick, 2013). Endogenous benzodiazepines exert a proconvulsant or partly an antiepileptic effect, because they can perform a positive or a negative allosteric modulation of the benzodiazepine-binding site at the $GABA_A$ receptor (Alho et al., 1991).

CONCLUSION

Benzodiazepines, which exert sedating and sleep-inducing, anxiolytic, muscle-relaxing, and antiepileptic effects, have been widely used since the 1960s. Benzodiazepines act on an allosteric site of the $GABA_A$ receptor located on diverse neurons. By exerting a positive allosteric modulatory action on the $GABA_A$ receptor, benzodiazepines enhance

GABAergic inhibition of dopaminergic and noradrenergic neurons in the brainstem, hippocampus, and prefrontal cortex. Benzodiazepines such as diazepam, lorazepam, and midazolam can be administered to treat acute seizures. However, when an acute seizure becomes a status epilepticus, these drugs lose their effectiveness, as a consequence of receptor trafficking. During a status epilepticus, $GABA_A$ receptors are internalized and NMDA glutamate receptors externalized. For this reason, it is recommended to combine a benzodiazepine (e.g., diazepam, lorazepam, or midazolam) with an NMDA receptor antagonist (such as ketamine) to treat successfully the status epilepticus. Updated from the previously published works, neural networks, including hypoactive $GABA_A$ GABAergic neurons and hyperactive NMDA glutamatergic neurons, have been described in the hippocampus in generalized epilepsy. Postsynaptic excitatory neurons, i.e., dopaminergic neurons and serotonergic neurons, have been included, while hyperactivity of D_2 dopaminergic neurons and hypoactivity of $5\text{-}HT_{2A}$ serotonergic neurons exert a proconvulsant effect. In these neural networks, a hyperactive action of A_{2A} adenosine neurons has been suggested in generalized epilepsy. It is important to know whether A_{2A} adenosine antagonists exert or not an antiepileptic effect as well as whether these drugs might be used as an add-on therapy. Endogenous benzodiazepines, the endozepines (DBI), and its cleavage products (TTN, ODN, and OD) are located in neurons and astrocytes. They are involved in ictogenesis in generalized epilepsy. They displace synthetic benzodiazepines from the benzodiazepine-binding site at the $GABA_A$ receptor, and at high doses, they can exert a positive allosteric modulation of this receptor. Consequently, they may exert either a proconvulsant or an anticonvulsant effect. The pharmacokinetic profiles, the application forms, and the indications of benzodiazepines have been described and compared. Diazepam, lorazepam, midazolam, and clonazepam can be used to treat acute seizures, but they should not be administered repeatedly because they might cause dependence, which is difficult to be cured. Clobazam can be administered to treat the Lennox–Gestaut syndrome in children, but its administration must be controlled due to its adverse effect. Besides the intravenous route, diazepam, lorazepam, and midazolam can also be administered by intranasal, buccal, or rectal routes. Finally, physicians should be careful in administering benzodiazepines to elderly patients, because they might have a paradoxical therapeutic effect.

ACKNOWLEDGMENTS

This work was supported by the "Programa XI: Financiación de Unidades de Excelencia de la *Universidad de Salamanca*" (Spain). The authors wish to thank Mr. Javier Muñoz (University of Seville, Spain) for technical assistance.

REFERENCES

Agrawal N., Mula M. 2019. Treatment of psychoses in patients with epilepsy: an update. *Therapeutic Advances in Psychopharmacology* 9: 1–10.

Alho H, Harjuntausta T, Schultz R, Pelto-Huikko M, Bovolin P. 1991. Immuno-histochemistry of diazepam-binding inhibitor (DBI) in the central nervous system and peripheral organs: its possible role as an endogenous regulator of different types of benzodiazepine receptors. *Neuropharmacology* 30: 1381–1386.

Butler J., Lewis M. 2001. Lorazepam or diazepam for generalized convulsions in adults. *Emergency Medicine Journal* 18: 116–123.

Butuzova M.V., Kitchigina V.E. 2008. Repeated blockade of $GABA_A$ receptors in the medial septal region induces epileptiform activity in the hippocampus. *Neuroscience Letters* 434: 133–138.

Calcaterra N.E., Barrow J.C. 2014. Classics in chemical neuroscience: diazepam (valium). *ACS Chemical Neuroscience* 5: 253–260.

Christian CA, Huguenard JR. 2013. Atrocyte potentiate GABAergic transmission in the thalamic reticular nucleus via endozepine signaling. *Proceedings of the National Academy of Sciences of the United States of America* 110: 20278–20283.

Clinckers R., Smolders I., Meurs A., Ebinger G., Michotte Y. 2004. Anticonvulsant action of GBR-12,909 and citalopram against acute experimentally induced limbic seizures. *Neuropharmacology* 47: 1053–1061.

Farzampour Z, Reimer RJ, Huguenhard J. 2015. Endozepines. *Advances in Pharmacology* 72: 147–164.

Ferrero P, Santi MR, Conti-Tronconi B, Costa E, Guidotti A. 1986. Study of an octadecaneuropeptide derived from diazepam-binding inhibitor (DBI): biological activity and presence in rat brain. *Proceedings of the National Academy of Sciences of the United States of America* 83: 827–831.

Heller A.H., Wargacki S., Jung C., Wyatt D.J., Schobel A.M. 2018. Safety and pharmacokinetics of diazepam buccal soluble film. *Neurology* 90 (suppl. 15): P4.272.

Hogan R.E., Gidal B.E., Koplowitz B., Koplowitz L.P., Lowenthal R.E., Carrazana E. 2020. Bioavailability and safety of diazepam intranasal solution compared to oral and rectal diazepam in healthy volunteers. *Epilepsia* 61: 455–464.

Huddart R., Leeder J.S., Altman R.B., Klein T.E. 2018. PharmGKB summary: clobazam pathway, pharmacokinetics. *Pharmacogenetics and Genomics* 28: 110–115.

Kasahara Y., Igata H., Sasaki T., Ikegaya Y., Koyama R. 2019. The pharmacological assessment of $GABA_A$ receptor activation in experimental febrile seizures in mice. *eNeuro* 6: ENEURO.0429–18.2019.

Knoester P.D., Jonker D.M., van der Hoeven R.T.M., et al. 2002. Pharmacokinetics and pharmacodynamics of midazolam administered as a concentrated intranasal spray. A study in healthy volunteers. *Clinical Pharmacology* 53: 501–507.

Knudsen J. 1991. Acyl-CoA binding and transport, an alternative function of diazepam-binding receptor (DBI), which is identical with acyl-CoA binding protein. *Neuropharmacology* 30: 1405–1410.

Kroll D.S., Nieva H.R., Barsky A.J., Linder J.A. 2016. Benzodiazepines are prescribed more frequently to patients already at risk for benzodiazepine-related adverse events in primary care. *Journal of General Internal Medicine* 31: 1027–1034.

Ku L.C., Hornik C.P., Beechinor R.J., et al. 2018. Population pharmacokinetics and exposure-response relationships of diazepam in children treated for status epilepticus. *CPT: Pharmcometrics & Systems Pharmacology* 7: 717–728.

Lian J.J., Cao Y.Q., Li Y.L., Yu G., Su R.B. 2020. Flumazenil-insensitive benzodiazepine effects in recombinant αβ and neuronal $GABA_A$ receptors. *Brain Sciences* 10: 150.

Loomis WF, Behrens MM, Williams ME, Anjard C. 2010. Pregnenolone sulfate and cortisol induce secretion of acyl-CoA binding protein and its conversion into endozepines from astrocytes. *The Journal of Biological Chemistry* 285: 21359–21365.

Mackway-Jones K. 2001. Towards evidence based emergency medicine: best BETSs from the Manchester royal infirmary. *Emergency Medicine Journal* 18: 116–123.

Niquet J., Baldwin R., Norman K., Suchomelova L., Lumley L., Wasterlain C.G. 2016. Midazolam-ketamine dual therapy stops cholinergic status epilepticus and reduces Morris water maze deficits. *Epilepsia* 57: 1406–1415.

Niquet J., Lumley L., Baldwin R., et al. 2019. Early polytherapy for benzodiazepine-refractory status epilepticus. *Epilepsy & Behavior* 101(Pt B): 106367.

Nitz I, Kruse ML, Klapper M, Döring F. 2011. Specific variation of low-abundance transcript variants encoding human Acyl-CoA binding protein (ACBP) isoforms. *Journal of Cellular and Molecular Medicine* 15: 909–927.

Porcello DM, Huntsmann MM, Mihalek RM, Homanics GE, Huguenard JR. 2003. Intact synaptic GABAergic inhibition and altered neurosteroid modulation of thalamic relay neurons in mice lacking delta subunit. *Journal of Neurophysiology* 89: 1378–1386.

Ramerstorfer J, Furtmüller R, Vogel E, Huck S, Sieghart W. 2010. The point mutation gamma 2F77I changes the potency and efficacy of benzodiazepine site ligands in different GABAA receptor subtypes. *European Journal of Pharmacology* 636: 18.27.

Rao S.K., Mahulikar A., Ibrahim M., Shah A., Seraji-Bozorgzad N., Mohamed W. 2018. Inadequate benzodiazepine dosing may result in progression to refractory and nonconvulsive status epilepticus. *Epileptic Disorders* 20: 265–269.

Reddy D.S., Kulkarni S.K. 1996. Role of GABA-A and mitochondrial diazepam binding inhibitor receptors in the anti-stress activity of neurosteroids in mice. *Psychopharmacology (Berlin)* 128: 280–292.

Rogawski M.A., Heller A.H. 2019. Diazepam buccal film for the treatment of acute seizures. *Epilepsy & Behavior* 101(Pt B): 106537.

Rye DB, Bliwise DL, Parker K, et al. 2012. Modulation of vigilance in the primary hypersomnias by endogenous enhancement of GABAA receptors. *Science Translational Medicine* 5(161): 161ra151.

Sánchez Fernández I., Goodkin H.P., Scott R.C. 2019. Pathophysiology of convulsive status epilepticus. *Seizure* 68: 16–21.

Sansalone L., Bratsch-Prince J., Tang S., Captain B., Mott D.D., Raymo F.M. 2019. Photopotentiation of the $GABA_A$ receptor with caged diazepam. *Proceedings of the National of Academy of Science of the United States of America* 116: 21176–21184.

Schols-Hendriks M.W., Lohmann J.J., Ranknegt R., Korten J.J., Markus F.W., Hooymans P.M. 1995. Absorption of clonazepam after intranasal and buccal administration. *British Journal of Clinical Pharmacology* 39: 449–451.

Sieghart W. 1990. GABA(A) benzodiazepine receptors and epilepsy. *Wiener Klinische Wochenschrift* 102: 197–201.

Slobodyansky E., Guidotti A., Wambebe C., Berkovich A., Costa E. 1989. Isolation and characterization of a rat brain triakontatetraneuropeptide, a post-translational product of diazepam-binding inhibitor. Specific action at the Re 5-4864 recognition site. *Journal of Neurochemistry* 53: 1276–1284.

Thomé-Souza M.S., Kuczynski E., Valente K.D. 2007. Sertraline and fluoxetine: safe treatment for children and adolescents with epilepsy and depression. *Epilepsy & Behavior* 10: 417–425.

Tölle R., Windgassen K. 2003. Tranquilizer/Anxiolytika. In: Rainer Tölle, Klaus Windgassen, Reinhart Lemmp and Reinmar du Bois (Eds.), *Psychiatrie*. Berlin: Springer, pp. 366–369.

Verdon B, Zheng J, Nicholson RA, Ganellin R, Lees G. 2000. Stereoselective modulatory actions on $GABA_A$ receptors and voltage-gated Na^+ channels *in vitro*: a putative endogenous ligand for depressant drugs in CNS. *British Journal of Pharmacology* 129: 283–290.

Wang H., Chen X., Jang J., Shi J., Hu P. 2016. Evaluating a physiologically based pharmacokinetic model for predicting the pharmacokineticcs of midazolam in Chinese after oral administration. *Acta Pharmacologica Sinica* 37: 276–284.

Werner F.M., Coveñas R. 2011. Classical neurotransmitters and neuropeptides involved in generalized epilepsy: focus on antiepileptic drugs. *Current Medicinal Chemistry* 13: 4933–4988.

Werner F.M., Coveñas R. 2015. Classical neurotransmitters and neuropeptides involved in generalized epilepsy: how can the antiepileptic effect be improved? *Journal of Neurology and Neurophysiology* 6: 4.

Werner F.M., Coveñas R. 2017a. Classical neurotransmitters and neuropeptides involved in generalized epilepsy in a multi-neurotransmitter system: how to improve the antiepileptic effect? *Epilepsy & Behavior* 71: 124–129.

Werner F.M., Coveñas R. 2017b. Influence of GABAergic Neurons and Other Neurotransmitters and Neuropeptides on Ictogenesis in Generalized Epilepsy. In: J. Stein (Ed.), *Reference Module in Neuroscience and Biobehavioral Psychology*. Oxford: Elsevier, pp. 1–6.

Werner F.M., Coveñas R. 2019. Neural networks in generalized epilepsies and novel antiepileptic drugs. *Current Pharmaceutical Design* 25: 1–5.

Wick J.Y. 2013. The history of benzodiazepines. *The Consultant Pharmacist* 28: 538–548.

Section IV

Implications for Tolerance, Withdrawal, and Abuse

14 Impact of Endogenous Benzodiazepines on Tolerance, Abuse, or Withdrawal

Robert B. Raffa
Temple University School of Pharmacy (Emeritus)
University of Arizona College of Pharmacy (Adjunct)
Neumentum Inc.
Enalare Therapeutics

Diana Amantea
University of Calabria

CONTENTS

TOLERANCE AND PHYSICAL DEPENDENCE

All drugs develop some degree of tolerance and physical dependence (two natural phenomena that are appropriate adaptations of the body in response to external manipulation—a type of compensatory feedback control process). The benzodiazepines (and "Z" drugs) are no exception. The phenomenon develops very rapidly for some drugs (it is then often termed "tachyphylaxis"); this phenomenon is so common for other drugs, and it is expected during the course of normal therapy (like with the opioids). By themselves, they are normal and generally not a concern, since they are the normal physiological adaptations to almost any drug. They are demonstrably not an independent indication of drug abuse. However, upon cessation of intake of the drug, physical dependence is manifested as withdrawal symptoms. These symptoms can be problematic for some patients no matter how short a time they were taking the drug. It can become even more problematic with longer exposure to the drug. Although the overwhelming majority of patients who properly take benzodiazepines are not current or future abusers, the development of tolerance and physical dependence can contribute to physiological cues that can make it uncomfortable and difficult to stop taking the drugs.

Tolerance to the effects of benzodiazepines mainly develops when they are used for a long time, requiring dose escalation and resulting in potential abuse (Licata and Rowlett, 2008). The majority of people take benzodiazepines following medical prescription, while they are misused by a small percentage of patients, whereby dose escalation and dependence may result in unintentional abuse (Schmitz, 2016). Thus, to avoid such complications, national guidelines from several countries recommend short-term (2–4 weeks) use of these medications at low therapeutic doses. Nevertheless, these indications are often underestimated and benzodiazepines are prescribed for long-term use, especially in the elderly and to obtain concessionary benefits (Lader, 2011). Notably, in less developed countries, medical prescription is not needed since benzodiazepines are sold over-the-counter, thus resulting in an increased frequency of overuse and abuse (Raoof et al., 2008).

Tolerance to benzodiazepines is configured as a form of neuronal plasticity characterized by different adaptive changes in the function of GABA-A receptors as a consequence of long-term use of these positive allosteric modulators. Depending on the pharmacological effect considered, tolerance develops over different time scales: relatively quickly for the sedative and hypnotic effects, followed by tolerance to the anticonvulsant action; however, for the anxiolytic and amnesic effects, some degree of tolerance has been shown to slowly develop in animal models, but in humans, it may be absent or partially develops after long-term use (Vinkers and Olivier, 2012). In fact, clinical studies show a persisting anxiolytic effect, at least for panic disorder, generalized anxiety disorder, and social phobia. By contrast, the development of tolerance during the first several months of treatment in a high percentage of epileptic patients limits the prescription of these drugs only for acute epileptic seizures or status epilepticus (Riss et al., 2008). Regarding tolerance to the sedative effects, although it is mainly reported after 1–2 weeks of treatment with benzodiazepines with a short half-life, the rate of elimination does not appear to be the only contributing factor (Soldatos et al., 1999). Indeed, pharmacokinetic factors do not seem to be relevant for the development of tolerance, being the GABA-A receptor the most obvious candidate (Cowley et al., 1995; Fernandes et al., 1996).

Tolerance and dependence are not simply associated with the down-regulation of receptor number, but involve complex mechanisms that depend on receptor subtypes and brain regions involved (Bateson, 2005; Vinkers and Olivier, 2012). The activation of distinct subtypes of the α-subunit strongly affects the development of tolerance. Accordingly, tolerance does not appear to occur with drugs selectively activating α2/α3-containing GABA-A receptor subtypes, whereas concomitant activation of α1- and α5-containing receptors is required for the development of tolerance to the sedative effects of diazepam (Van Rijnsoever et al., 2004; Vinkers et al., 2012). Intriguingly, nonselective low-affinity partial agonists at the α-subunit are less prone to develop tolerance, thus representing promising antiepileptic drugs (Rundfeldt et al., 1995; Rundfeldt and Löscher, 2014).

Prolonged treatment with benzodiazepines may induce a number of adaptive changes that directly or indirectly affect the GABAergic synapse, including GABA-A receptor uncoupling; alterations in receptor subunits expression; altered intracellular receptor trafficking; changes in signal transduction pathways; compensatory glutamate sensitization; and modulation of serotonin, dopamine, and acetylcholine systems (Vinkers and Olivier, 2012; Lorenz-Guertin et al., 2019). Moreover, there is evidence that a dynamic interaction exists between the neurosteroid system and the GABA-A receptor, exerting a reciprocal influence on the development of tolerance (Reddy and Kulkarni, 1997; Herd et al., 2007; Carver and Reddy, 2016).

GABA-A receptor uncoupling results in a decreased ability of benzodiazepines to enhance GABA-induced inhibitory postsynaptic potential and may be triggered by changes in receptor subunits composition, alterations of the receptor status (e.g., phosphorylation), reduced ability to activate second messengers, or conformational changes (Vinkers et al., 2012; Gravielle, 2016).

A decreased coupling may develop as a result of changes in GABA-A receptor subunit composition, modifications (e.g., phosphorylation) of the receptor itself or its second messengers, or any process affecting the conformational state of the receptor (Primus et al., 1996; Ali and Olsen, 2001; Gravielle et al., 2005; Gutiérrez et al., 2014; Ferreri et al., 2015). Notably, this is an aspecific process, as it can be induced by compounds acting on different modulatory sites of the GABA-A receptor, including barbiturates and neurosteroids (Friedman et al., 1996). This adds further complexity to the development of tolerance.

Thus, although benzodiazepines were widely used in the past to treat anxiety conditions, they are no longer recommended as first-line therapies because of the risks associated with their chronic use (Ravindran and Stein, 2010). Thus, while many clinicians still consider benzodiazepines for both acute anxiety and chronic anxiety, they are actually very effective against acute episodes, but when taken for a long time, especially at high doses, they may result in disadvantageous adverse effects, including physiological and psychological dependence, impaired cognition and coordination, and potential fatalities upon withdrawal, among others.

WITHDRAWAL

Withdrawal symptoms normally reveal the unopposed compensatory mechanism, opposite to the drug-induced effects. Heightened anxiety, therefore, is a withdrawal symptom that is expected with anxiolytic drugs. However, withdrawal from benzodiazepines can give rise to a curious array of symptoms that are often difficult to understand as opposite to benzodiazepine-induced effects (Table 14.1).

Thus, in addition to an increased anxiety, the abrupt interruption of long-term benzodiazepine treatment may result in diverse symptoms, including insomnia and sensory disturbances, that are determining factors for dependence (Allison and Pratt, 2003; O'Brien, 2005). These symptoms may persist for 1–2 weeks after drug discontinuation and are the consequence of neuronal adaptations occurring during chronic benzodiazepine treatment (Allison and Pratt, 2003). However, some uncomfortable symptoms may last for a paradoxically long time, which is not always correlated with the time necessary to eliminate the drug from the body, neither with the dose nor with the treatment duration.

Possible mechanisms involved in the development of withdrawal include modifications of pharmacodynamic (i.e., long-term alterations of receptor number and

TABLE 14.1
Withdrawal Symptoms (Immediate or Delayed and Protracted) Reported upon Interruption of Benzodiazepine Use

Anxiety/panic attacks	Tremors
Insomnia	Tinnitus
Headaches	Paresthesia
Constant flu symptoms	Unusual fatigue, weakness
Sweating	Nausea
Loss of appetite	Pain in neck and shoulders
Restless legs	Hair loss
Sensory fog	Depersonalization
Muscle pain	Hallucinations
Delirium	Vertigo
Muscle spasms	Anhedonia
Agoraphobia	Intrusive thoughts
Hyper sensory sensitivity	GI: constipation or diarrhea

function) and/or pharmacokinetic properties. For example, rebound insomnia is mainly reported with benzodiazepines with a short half-life, although the rate of elimination does not appear to be the only contributing factor (Soldatos et al., 1999).

Being positive allosteric modulators of the GABA-A receptor, benzodiazepines may reduce receptor efficiency after long-term use, thus leading to tolerance. Conversely, when treatment is abruptly interrupted, the down-regulated state of inhibitory neurotransmission is unmasked. Although tolerance and withdrawal may arise from compensatory mechanisms, tolerance may develop in the absence of dependence, while the appearance of withdrawal symptoms is not necessarily associated with tolerance (Licata and Rowlett, 2008). Accordingly, withdrawal symptoms may arise upon abrupt discontinuation of compounds, such as zolpidem and clorazepate, that are not normally associated with obvious tolerance (Scherkl et al., 1989; Griffiths et al., 1992; Griffiths and Weerts, 1997).

Several studies have evaluated the molecular mechanisms underlying withdrawal after (sub)chronic treatment with benzodiazepines, leading to the conclusion that $\alpha2/\alpha3$ subtype selective compounds do not produce withdrawal symptoms, although abuse liability may still be present (Scherkl et al., 1989; Griffiths and Weerts, 1997; Vinkers et al., 2012). Nevertheless, the role of $\alpha1$-containing GABA-A receptors in mediating the development of physical dependence depends on the symptom studied (e.g., anxiety, seizures, etc) and on treatment duration (Kovačević et al., 2014). Moreover, lack of activity at the $\alpha1$-subunit is not sufficient to eliminate the appearance of withdrawal symptoms upon the discontinuation of benzodiazepines (Kovačević et al., 2014). Similar to other positive modulators, changes in the subunit composition of GABA-A receptor have also been reported to occur during benzodiazepine withdrawal (Uusi-Oukari and Korpi, 2010). Withdrawal of diazepam or other positive modulators of GABA-A receptor, such as ethanol and neurosteroids, is associated with a marked increase in the abundance of the α4-subunit mRNA that contributes to changes in receptor sensitivity and function (Biggio et al., 2003).

Also, glutamatergic sensitization, involved in the development of tolerance (as stated above), may underlie withdrawal symptoms after chronic benzodiazepine discontinuation. In fact, the persistent stimulation of the inhibitory GABAergic tone may trigger a compensatory increase in excitatory neurotransmission, causing rebound anxiety and seizure activity upon treatment interruption (Stephens, 1995; Allison and Pratt, 2003).

Curiously, benzodiazepine withdrawal symptoms in some patients persist a perplexingly long time, long after the last dose was taken and long after the drug has seemingly been eliminated from the body. The duration of these symptoms is not always correlated with dose level or duration of use. There is currently no satisfying understanding or mechanistic explanation for the protracted withdrawal symptoms in some patients, particularly since such a diverse spectrum of symptoms is displayed, and because not every patient experiences the same symptoms. Some hypothetical possibilities are long-term alterations in benzodiazepine receptor (or other receptor) number (up- or down-regulation), alteration of promoters of benzodiazepine receptor protein synthesis or degradation, alteration in ADME (absorption, distribution, metabolism, elimination), or a change in $GABA_A$ receptor function or subtype distribution, or possibly some involvement of peripheral benzodiazepine-binding sites.

IMPACT OF ENDOGENOUS BENZODIAZEPINES

Endogenous ligands at the GABA-A receptor may affect the development of tolerance and withdrawal; in turn, long-lasting exposure to positive modulators of GABA-A receptor, as well as their withdrawal, may affect the receptor structure and function altering its response to endogenous ligands, such as endozepines and neurosteroids (Biggio et al., 2003). Several endozepines, including diazepam-binding inhibitor (DBI) and octadecaneuropeptide $DBI_{(33-50)}$ (ODN), compete with the binding of the anxiolytic drug diazepam (Costa and Guidotti, 1991).

Intriguingly, withdrawal of a number of abused substances, including ethanol, nicotine, and morphine, causes an elevation of brain DBI mRNA levels (Ohkuma et al., 2001). In turn, the DBI-derived peptide ODN induces anxiety behavior in experimental animals, whereas i.c.v. injection of low doses of DBI in rabbits elicits convulsive and hyperactive responses, resembling withdrawal (Tonon et al., 2020). DBI and its fragments also bind to the peripheral benzodiazepine receptor to stimulate cholesterol transport into mitochondria, thus increasing the concentration of this metabolite, which is rate limiting for the synthesis of neurosteroids (Papadopoulos et al., 1991; Korneyev et al., 1993; Costa et al., 1994). By binding to an allosteric site on the GABA-A receptor, distinct from the benzodiazepine-binding site, neurosteroids potentiate synaptic and extrasynaptic GABA-A receptor functions. Thus, progesterone, through conversion to its neurosteroid metabolite 3α-OH-5α-pregnan-20-one (3α,5α-THP) in the brain, potentiates GABA responses in a benzodiazepine-like manner. Therefore, reduced circulating levels of progesterone and, consequently, of brain levels of 3α,5α-THP are associated with symptoms of premenstrual syndrome

(PMS), such as anxiety and seizure susceptibility, which are similar to benzodiazepine withdrawal symptoms. Conversely, withdrawal of progesterone increases seizure susceptibility and may result in insensitivity to benzodiazepine sedatives by enhancing the transcription of the gene encoding the α4-subunit of the GABA-A receptor (Moran et al., 1998; Smith et al., 1998). Another example is provided by the evidence that tolerance may be promoted during the menstrual cycle and pregnancy by continuous stimulation by the endogenous ligand allopregnanolone through selective alterations in GABA-A receptor subunit expression (Turkmen et al., 2011).

Figure 14.1 illustrates the error that is incurred in thinking about benzodiazepine administration as an independent process. If the level (concentration) of endogenous benzodiazepines is not taken into account, the concentration of benzodiazepine at the receptor site is erroneously assumed to be the same as the exogenous (administered) benzodiazepine that reaches the site. The observed effect is then erroneously

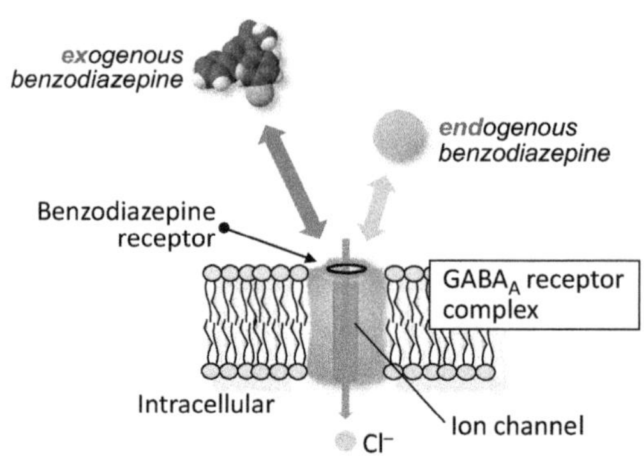

$$(B_Z + B_{endog}) + R_B \leftrightarrow B_ZR_B$$

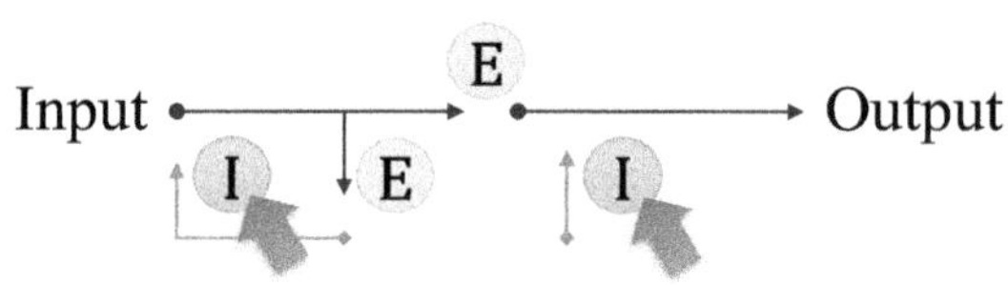

FIGURE 14.1 The usual depiction of a drug–receptor interaction includes only the exogenous drug molecules. This is usually sufficient for radioligand-binding studies and isolated tissue preparations, but even in these situations, inaccuracies can arise. In order to accurately understand the clinical response to a drug (therapeutic effect or adverse effects), the total contribution of the concentration of exogenous molecules (B_Z) and concentration of endogenous molecules (B_{endog}) needs to be considered (equation and arrows in control-system graph). Although the amount of the endogenous dose is a constant—per product labeling—the amount of endogenous substance is highly patient-specific (distributed on a bell-shaped curve). Thus, the clinical response to the same drug is highly individualized and highly variable from patient to patient. The same is true during cessation of treatment regarding the nature, magnitude, and duration of withdrawal.

assumed to result solely from the administered amount of drug. A relationship between the administered drug and the observed response (the classic drug-response relationship) is then erroneously assumed to be due exclusively to administered drug. But the presence of endogenous benzodiazepines changes all this. In actuality, the administered amount of benzodiazepine enters a system in which some basal level of endogenous benzodiazepine is already present. At the very least, this affects the administered drug's "apparent" dose–response relationship. To be an accurate equation, the already present endogenous benzodiazepine has to be added to the amount of administered drug.

Importantly, although the amount of exogenous benzodiazepine administered (added) might be the same for each patient, the amount of endogenous benzodiazepine could be significantly different for each patient, due to polymorphisms or other factors affecting ADME (absorption, distribution, metabolism, elimination). Indeed, one would assume that there would be a normal distribution of concentration of endogenous substance, and therefore, the sum of exogenous plus endogenous benzodiazepine would also follow a normal distribution—not the seemingly uniform amount of administered dose. Given a normal distribution, most patients would be somewhere in the middle region of the curve, but individual patients will be at the high or low end. Therefore, a particular patient might:

- respond as expected to a recommended dose of benzodiazepine (per the product label)
- appear to be a "nonresponder" if at the low end of the bell-shaped curve
- have exaggerated effects or exaggerated adverse effects if at the high end of the bell-shaped curve

The problems would not be easily recognized by the healthcare provider who forgets about the fact that therapy is added on to an existing basal level of benzodiazepine activity. And even if recognized, it would be unpredictable prior to therapy.

The same situation applies to cessation of therapy (rapid or tapered withdrawal). The amount of physical dependence developed through the course of therapy differs significantly from patient to patient because of the variability in the basal level of endogenous benzodiazepine (despite the same administered dose). This would manifest itself as a large interpatient variability in the nature, and duration, of withdrawal signs—minimal and short term for some patients, greatly exaggerated and protracted (and seemingly perplexing) for other patients.

REFERENCES

Ali NJ, Olsen RW (2001) Chronic benzodiazepine treatment of cells expressing recombinant GABAA receptors uncouples allosteric binding: Studies on possible mechanisms. *J Neurochem* 79:1100–1108.

Allison C, Pratt JA (2003) Neuroadaptive processes in GABAergic and glutamatergic systems in benzodiazepine dependence. *Pharmacol Ther* 98:171–195.

Bateson A (2005) Basic pharmacologic mechanisms involved in benzodiazepine tolerance and withdrawal. *Curr Pharm Des* 8:5–21.

Biggio G, Dazzi L, Biggio F, Mancuso L, Talani G, Busonero F, Mostallino MC, Sanna E, Follesa P (2003) Molecular mechanisms of tolerance to and withdrawal of GABAA receptor modulators. *Eur Neuropsychopharmacol* 13:411–423.

Carver CM, Reddy DS (2016) Neurosteroid structure-activity relationships for functional activation of extrasynaptic δGABAA receptors. *J Pharmacol Exp Ther* 357:188–204.

Costa E, Cheney DL, Grayson DR, Korneyev A, Longone P, Pani L, Romeo E, Zivkovich E, Guidotti A (1994) Pharmacology of neurosteroid biosynthesis. Role of the mitochondrial DBI receptor (MDR) complex. *Ann N Y Acad Sci* 746:223–242.

Costa E, Guidotti A (1991) Diazepam binding inhibitor (DBI): A peptide with multiple biological actions. *Life Sci* 49:325–344.

Cowley DS, Roy-Byme PP, Radant A, Ritchie JC, Greenblatt DJ, Nemeroff CB, Hommer DW (1995) Benzodiazepine sensitivity in panic disorder: Effects of chronic alprazolam treatment. *Neuropsychopharmacology* 12:147–157.

Fernandes C, File SE, Berry D (1996) Evidence against oppositional and pharmacokinetic mechanisms of tolerance to diazepam's sedative effects. *Brain Res* 734:236–242.

Ferreri MC, Gutiérrez ML, Gravielle MC (2015) Tolerance to the sedative and anxiolytic effects of diazepam is associated with different alterations of GABAA receptors in rat cerebral cortex. *Neuroscience* 310:152–162.

Friedman LK, Gibbs TT, Farb DH (1996) γ-Aminobutyric acidA receptor regulation: Heterologous uncoupling of modulatory site interactions induced by chronic steroid, barbiturate, benzodiazepine, or GABA treatment in culture. *Brain Res* 707:100–109.

Gravielle MC (2016) Activation-induced regulation of GABAA receptors: Is there a link with the molecular basis of benzodiazepine tolerance? *Pharmacol Res* 109:92–100.

Gravielle MC, Faris R, Russek SJ, Farb DH (2005) GABA induces activity dependent delayed-onset uncoupling of GABA/benzodiazepine site interactions in neocortical neurons. *J Biol Chem* 280:20954–20960.

Griffiths RR, Sannerud CA, Ator NA, Brady J V. (1992) Zolpidem behavioral pharmacology in baboons: Self-injection, discrimination, tolerance and withdrawal. *J Pharmacol Exp Ther* 260:1199–1208.

Griffiths RR, Weerts EM (1997) Benzodiazepine self-administration in humans and laboratory animals - Implications for problems of long-term use and abuse. *Psychopharmacology (Berl)* 134:1–37.

Gutiérrez ML, Ferreri MC, Farb DH, Gravielle MC (2014) GABA-induced uncoupling of GABA/benzodiazepine site interactions is associated with increased phosphorylation of the GABAA receptor. *J Neurosci Res* 92:1054–1061.

Herd MB, Belelli D, Lambert JJ (2007) Neurosteroid modulation of synaptic and extrasynaptic GABAA receptors. *Pharmacol Ther* 116:20–34.

Korneyev A, Pan BS, Polo A, Romeo E, Guidotti A, Costa E (1993) Stimulation of brain pregnenolone synthesis by mitochondrial diazepam binding inhibitor receptor ligands in vivo. *J Neurochem* 61:1515–1524.

Kovačević J, Timić T, Tiruveedhula VV., Batinić B, Namjoshi OA, Milić M, Joksimović S, Cook JM, Savić MM (2014) Duration of treatment and activation of α1-containing GABAA receptors variably affect the level of anxiety and seizure susceptibility after diazepam withdrawal in rats. *Brain Res Bull* 104:1–6.

Lader M (2011) Benzodiazepines revisited-will we ever learn? *Addiction* 106:2086–2109.

Licata SC, Rowlett JK (2008) Abuse and dependence liability of benzodiazepine-type drugs: GABAA receptor modulation and beyond. *Pharmacol Biochem Behav* 90:74–89.

Lorenz-Guertin JM, Bambino MJ, Das S, Weintraub ST, Jacob TC (2019) Diazepam accelerates GABAAR synaptic exchange and alters intracellular trafficking. *Front Cell Neurosci* 13:163.

Moran MH, Goldberg M, Smith SS (1998) Progesterone withdrawal II: Insensitivity to the sedative effects of a benzodiazepine. *Brain Res* 807:91–100.

O'Brien CP (2005) Benzodiazepine use, abuse, and dependence. *J Clin Psychiatry* 66:28–33.

Ohkuma S, Katsura M, Tsujimura A (2001) Alterations in cerebral diazepam binding inhibitor expression in drug dependence: A possible biochemical alteration common to drug dependence. *Life Sci* 68:1215–1222.

Papadopoulos V, Berkovich A, Krueger KE, Costa E, Guidotti A (1991) Diazepam binding inhibitor and its processing products stimulate mitochondrial steroid biosynthesis via an interaction with mitochondrial benzodiazepine receptor. *Endocrinology* 129:1481–1488.

Primus RJ, Yu J, Xu J, Hartnett C, Meyyappan M, Kostas C, Ramabhadran T V., Gallager DW (1996) Allosteric uncoupling after chronic benzodiazepine exposure of recombinant γ-Aminobutyric AcidA receptors expressed in Sf9 cells: Ligand efficacy and subtype selectivity. *J Pharmacol Exp Ther* 276:882–890.

Raoof M, Nawaz H, Nusrat R, Pabaney AH, Randhawa AR, Rehman R, Rizvi NB, Naqvi H (2008) Awareness and use of benzodiazepines in healthy volunteers and ambulatory patients visiting a tertiary care hospital: A cross sectional survey Heo M, ed. *PLoS One* 3:e1804.

Ravindran LN, Stein MB (2010) The pharmacologic treatment of anxiety disorders: A review of progress. *J Clin Psychiatry* 71:839–854.

Reddy DS, Kulkarni SK (1997) Neurosteroid coadministration prevents development of tolerance and augments recovery from benzodiazepine withdrawal anxiety and hyperactivity in mice. *Methods Find Exp Clin Pharmacol* 19:395–405.

Riss J, Cloyd J, Gates J, Collins S (2008) Benzodiazepines in epilepsy: Pharmacology and pharmacokinetics. *Acta Neurol Scand* 118:69–86.

Rundfeldt C, Löscher W (2014) The pharmacology of imepitoin: The first partial benzodiazepine receptor agonist developed for the treatment of epilepsy. *CNS Drugs* 28:29–43.

Rundfeldt C, Wlaz P, Honack D, Loscher W (1995) Anticonvulsant tolerance and withdrawal characteristics of benzodiazepine receptor ligands in different seizure models in mice. Comparison of diazepam, Bretazenil and Abecarnil. *J Pharmacol Exp Ther* 275:693–702.

Scherkl R, Kurudi D, Frey HH (1989) Clorazepate in dogs: Tolerance to the anticonvulsant effect and signs of physical dependence. *Epilepsy Res* 3:144–150.

Schmitz A (2016) Benzodiazepine use, misuse, and abuse: A review. *Ment Heal Clin* 6:120–126.

Smith SS, Gong QH, Hsu FC, Markowitz RS, Ffrench-Mullen JMH, Li X (1998) GABA(A) receptor α4 subunit suppression prevents withdrawal properties of an endogenous steroid. *Nature* 392:926–930.

Soldatos CR, Dikeos DG, Whitehead A (1999) Tolerance and rebound insomnia with rapidly eliminated hypnotics: A meta-analysis of sleep laboratory studies. *Int Clin Psychopharmacol* 14:287–303.

Stephens DN (1995) A glutamatergic hypothesis of drug dependence: Extrapolations from benzodiazepine receptor ligands. *Behav Pharmacol* 6:425–446.

Tonon M-C, Vaudry H, Chuquet J, Guillebaud F, Fan J, Masmoudi-Kouki O, Vaudry D, Lanfray D, Morin F, Prevot V, Papadopoulos V, Troadec J-D, Leprince J (2020) Endozepines and their receptors: Structure, functions and pathophysiological significance. *Pharmacol Ther* 208:107386.

Turkmen S, Backstrom T, Wahlstrom G, Andreen L, Johansson IM (2011) Tolerance to allopregnanolone with focus on the GABA-A receptor. *Br J Pharmacol* 162:311–327.

Uusi-Oukari M, Korpi ER (2010) Regulation of GABAA receptor subunit expression by pharmacological agents. *Pharmacol Rev* 62:97–135.

Van Rijnsoever C, Täuber M, Choulli MK, Keist R, Rudolph U, Mohler H, Fritschy JM, Crestani F (2004) Requirement of α5-GABAA receptors for the development of tolerance to the sedative action of diazepam in mice. *J Neurosci* 24:6785–6790.

Vinkers CH, Olivier B (2012) Mechanisms underlying tolerance after long-term benzodiazepine use: A future for subtype-selective receptor modulators? *Adv Pharmacol Sci* 2012:1–19.

Vinkers CH, van Oorschot R, Nielsen EØ, Cook JM, Hansen HH, Groenink L, Olivier B, Mirza NR (2012) GABAA receptor α subunits differentially contribute to diazepam tolerance after chronic treatment Homberg J, ed. *PLoS One* 7:e43054.

Index

Note: **Bold** page numbers refer to tables and *Italic* page numbers refer to figures.

For Product Safety Concerns and Information please contact our EU representative GPSR@taylorandfrancis.com
Taylor & Francis Verlag GmbH, Kaufingerstraße 24, 80331 München, Germany

www.ingramcontent.com/pod-product-compliance
Ingram Content Group UK Ltd.
Pitfield, Milton Keynes, MK11 3LW, UK
UKHW020202200926
475702UK00004B/577
9781032043104